Florian Dombois (ed.):
The Wind Tunnel Model — Transdisciplinary Encounters

Scheidegger & Spiess

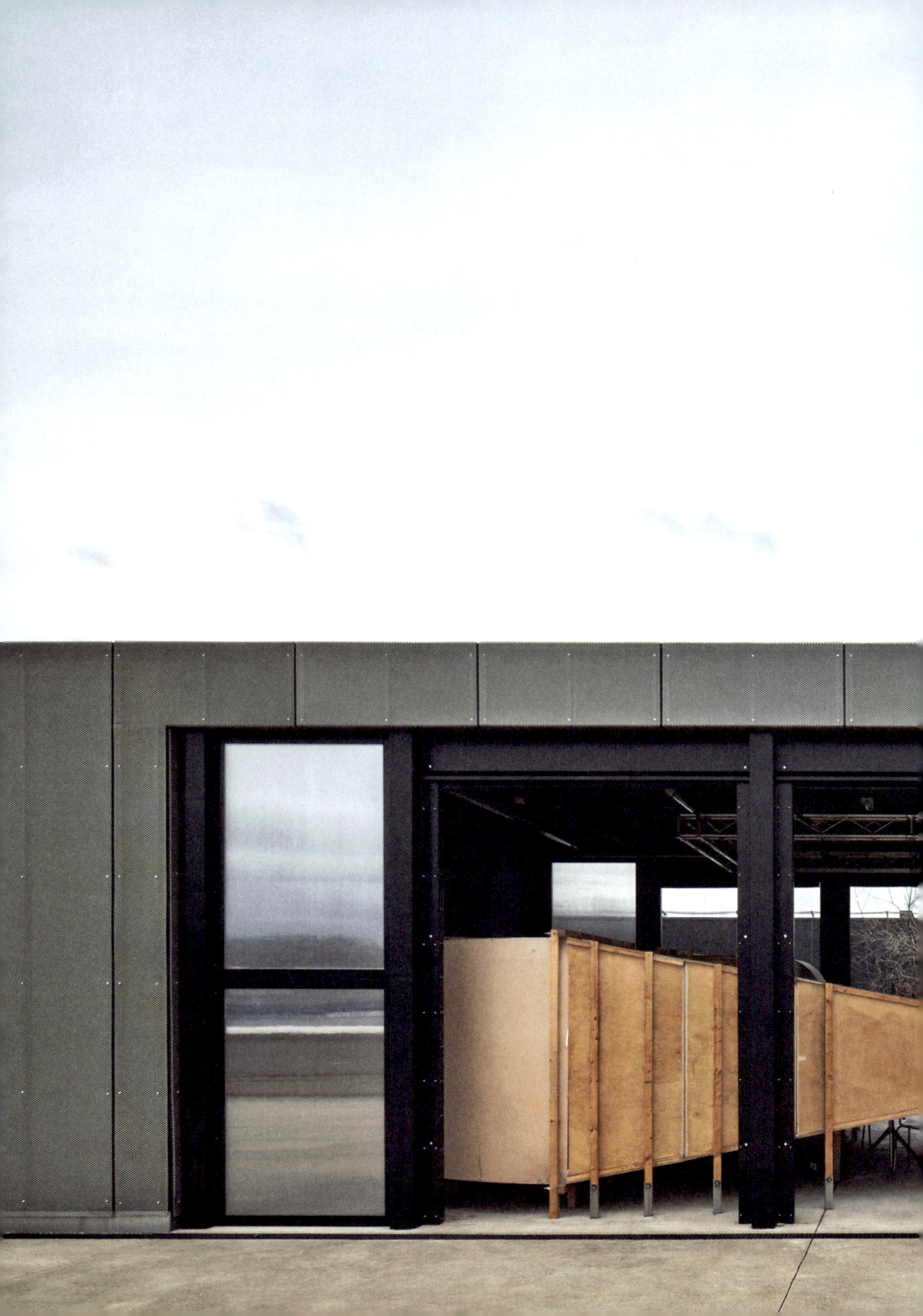

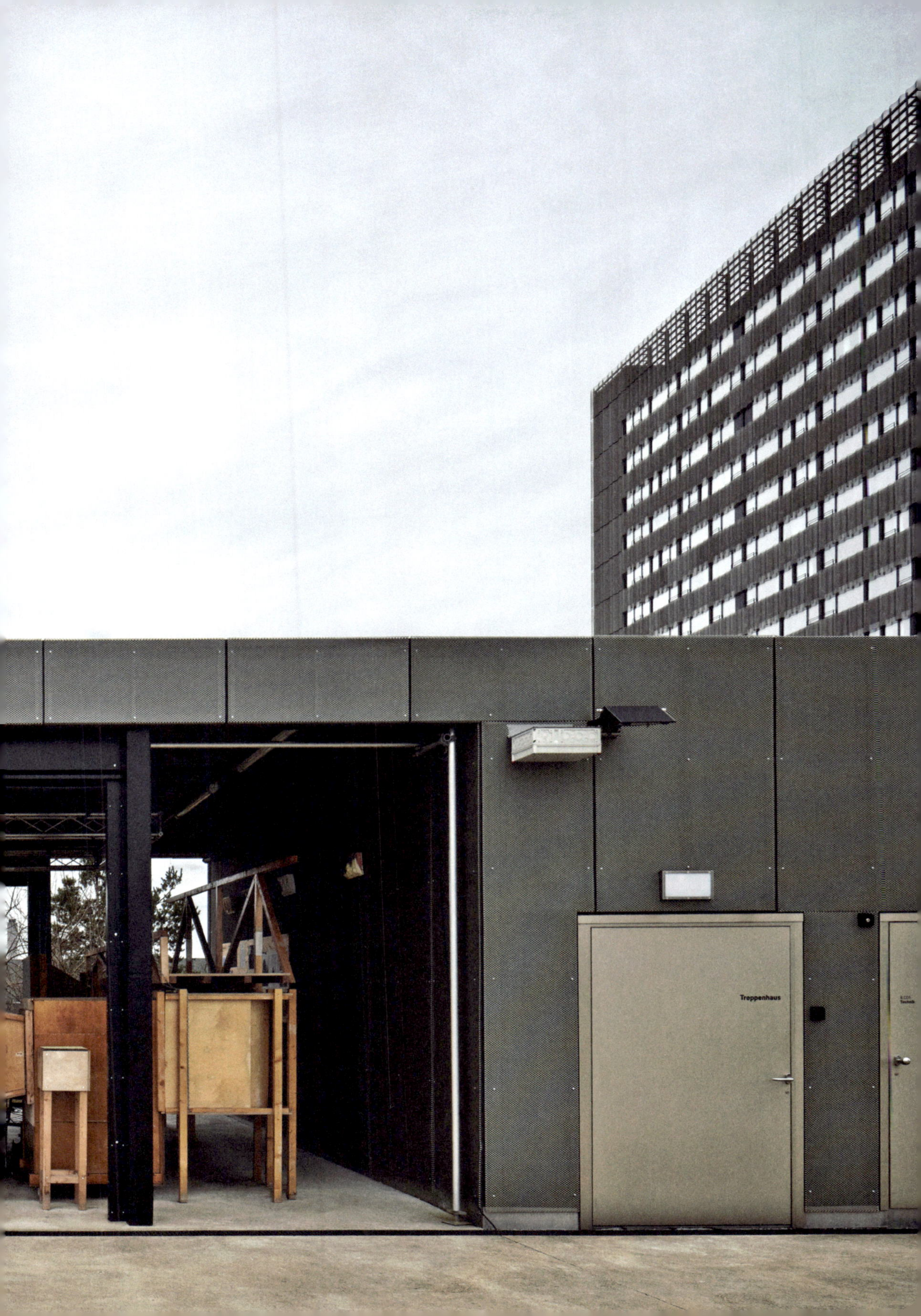
Treppenhaus

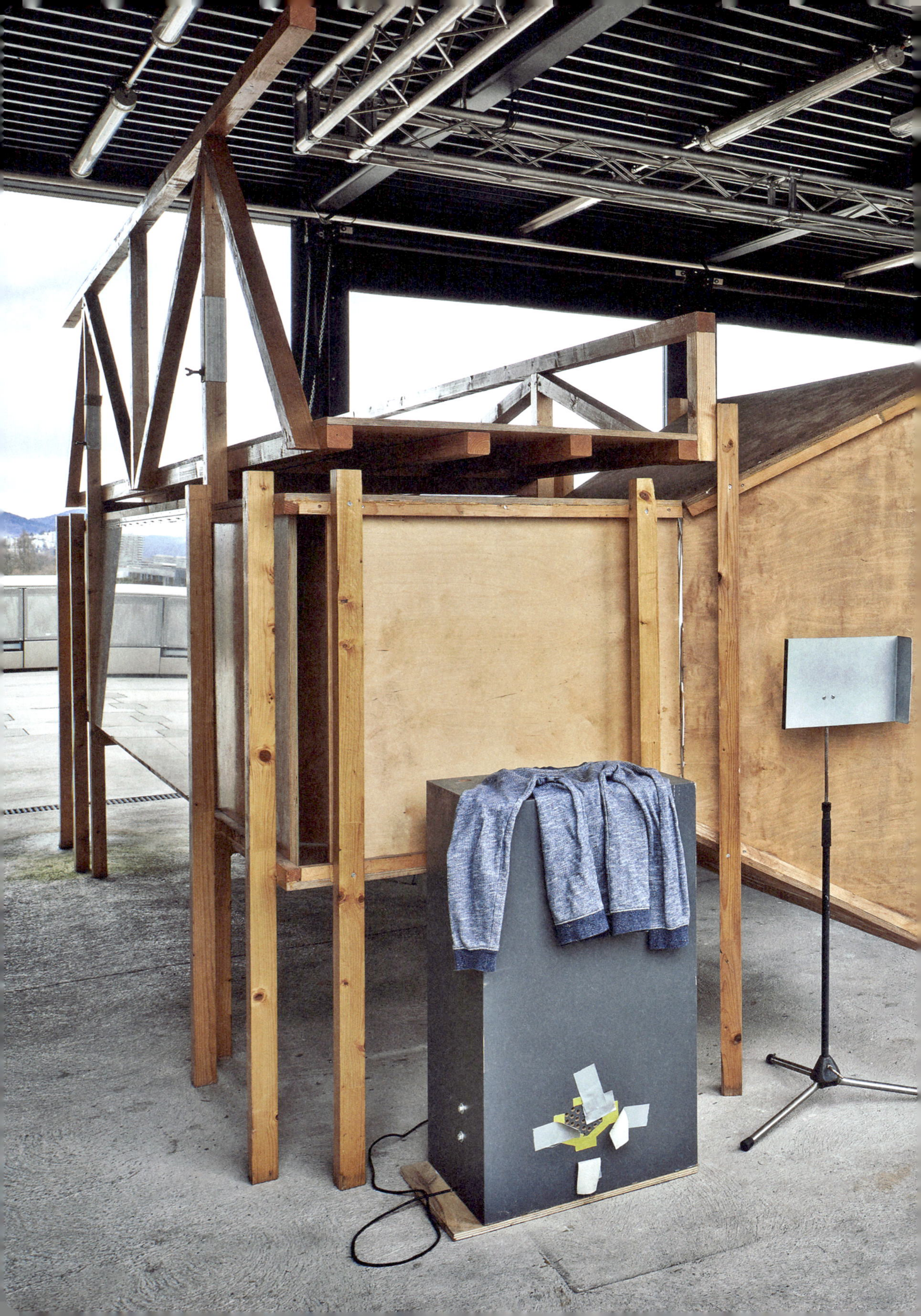

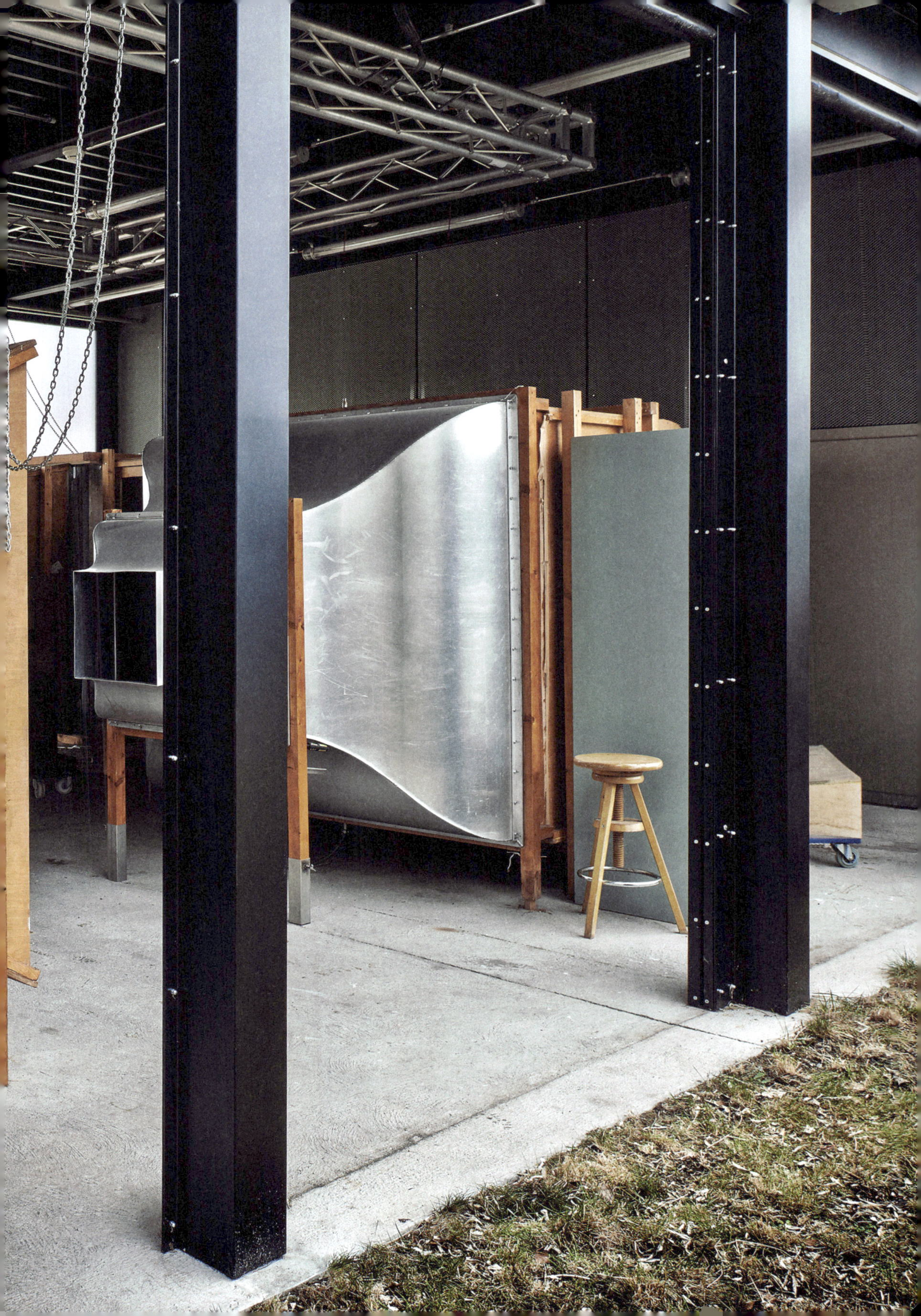

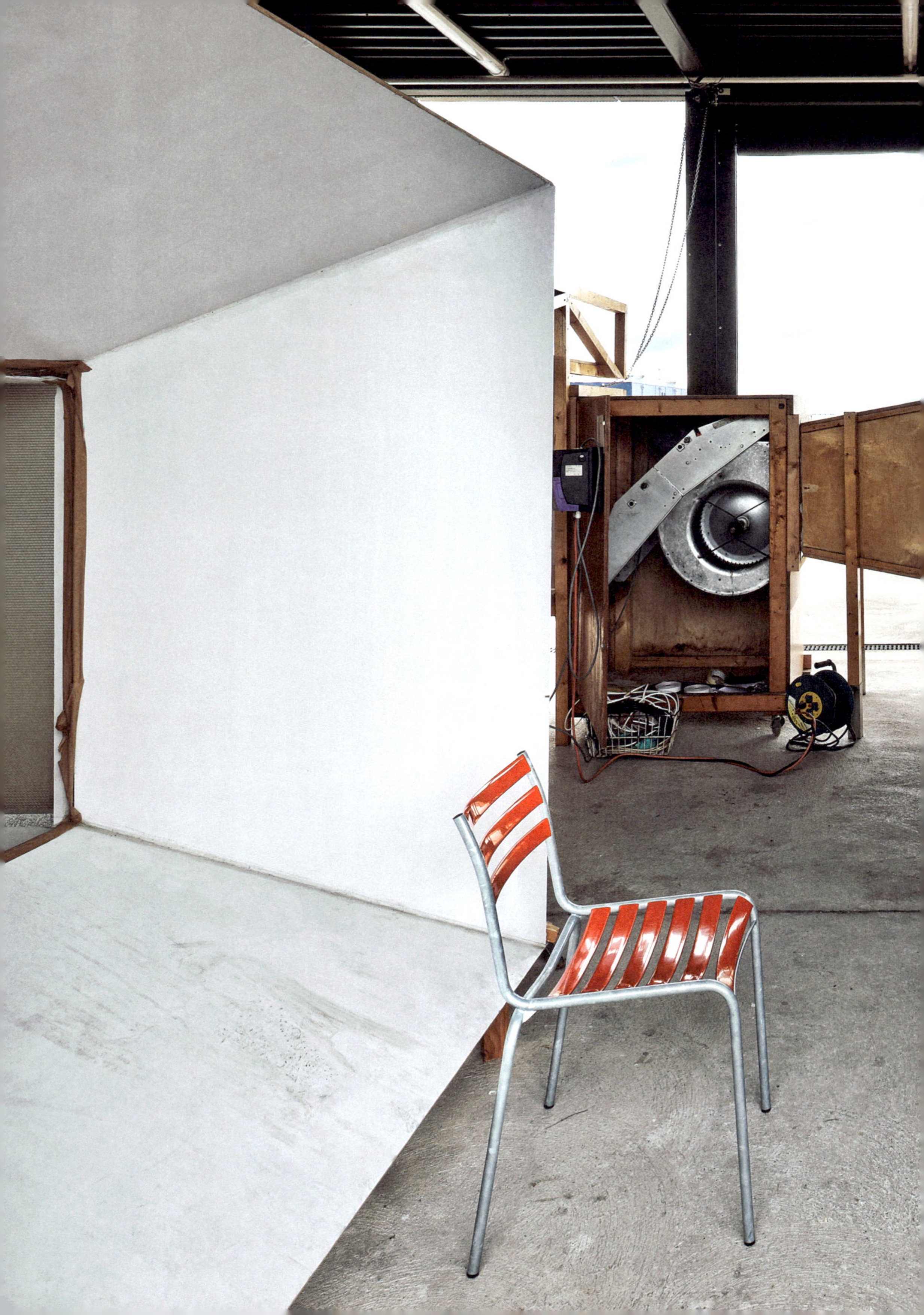

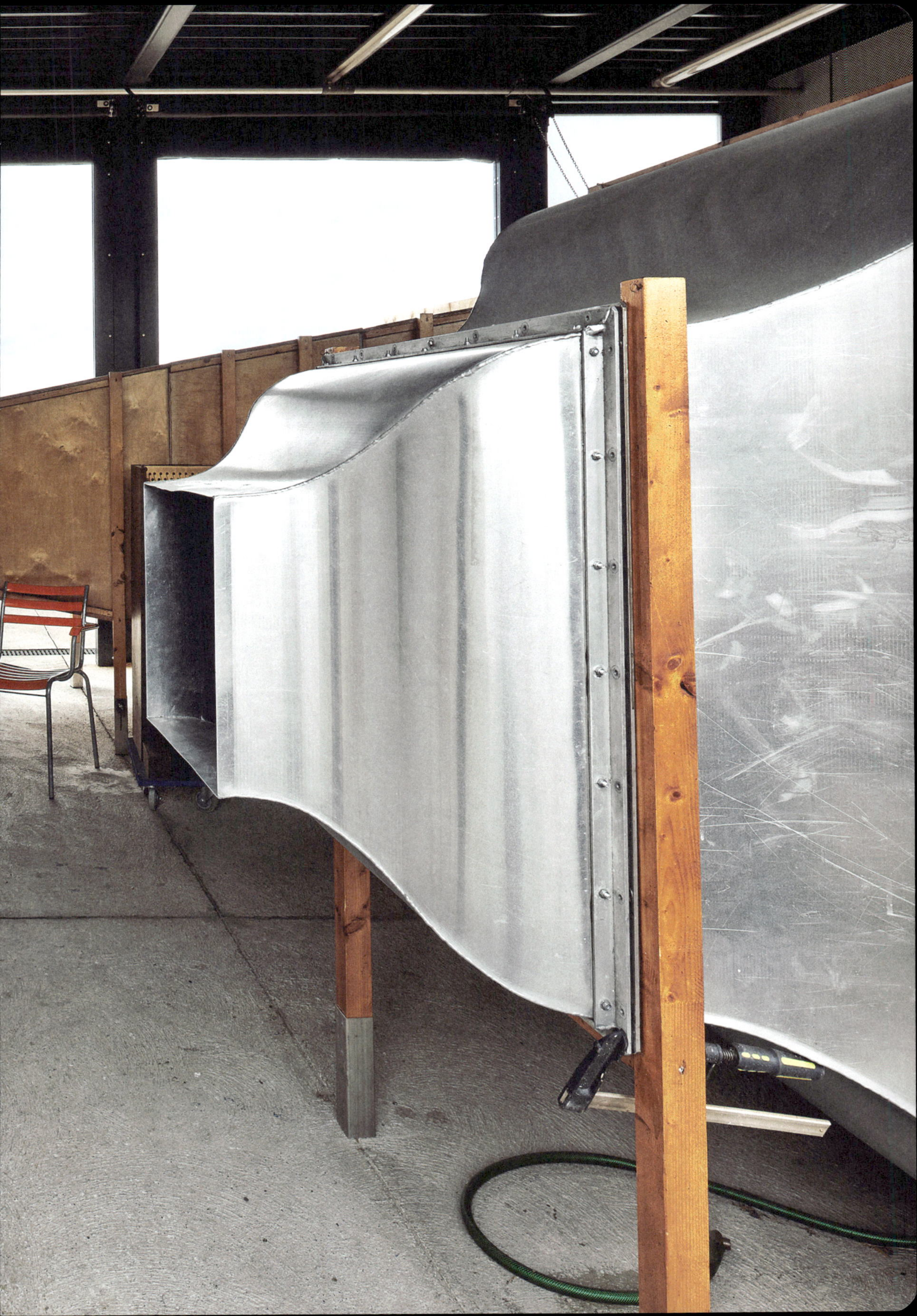

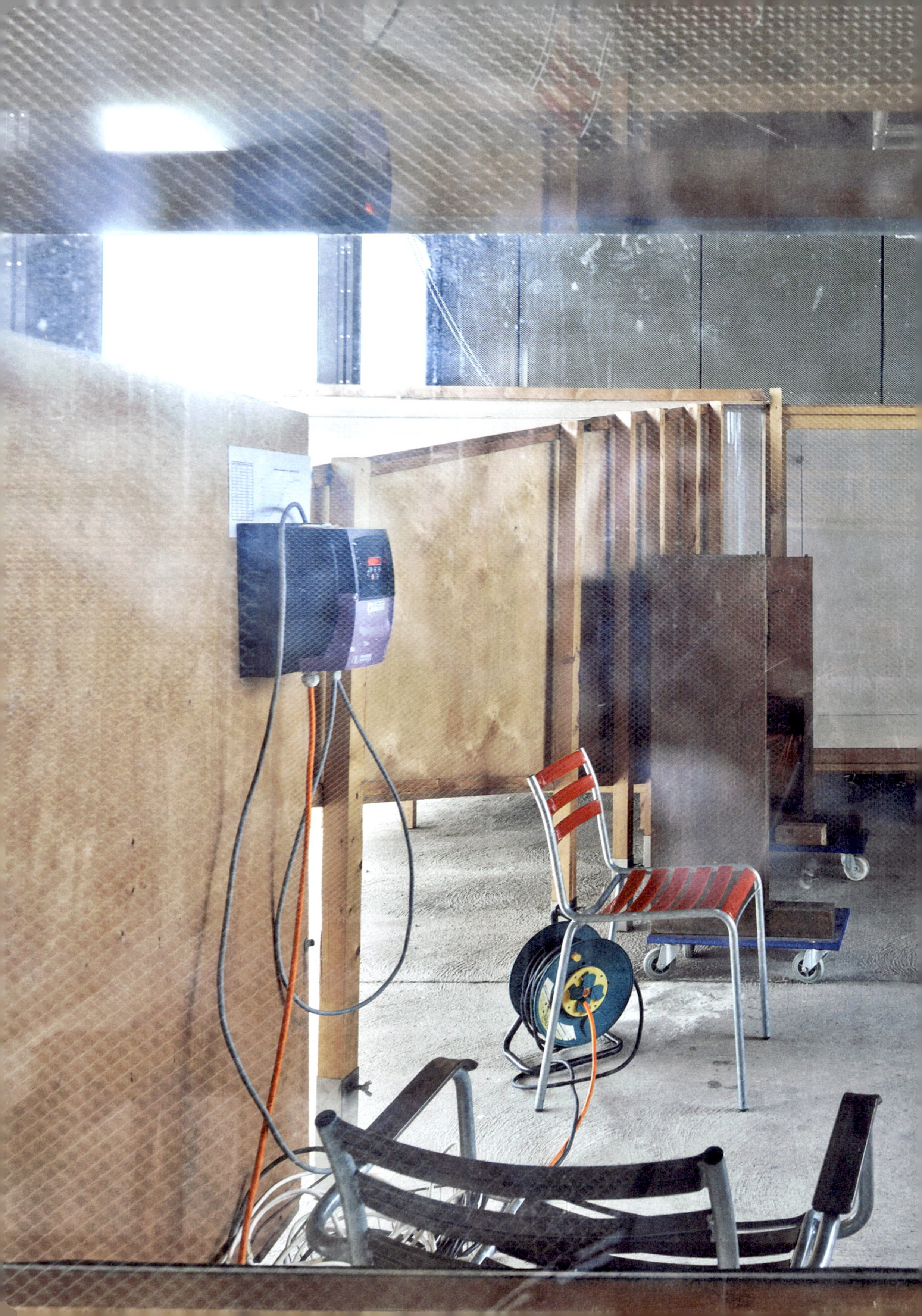

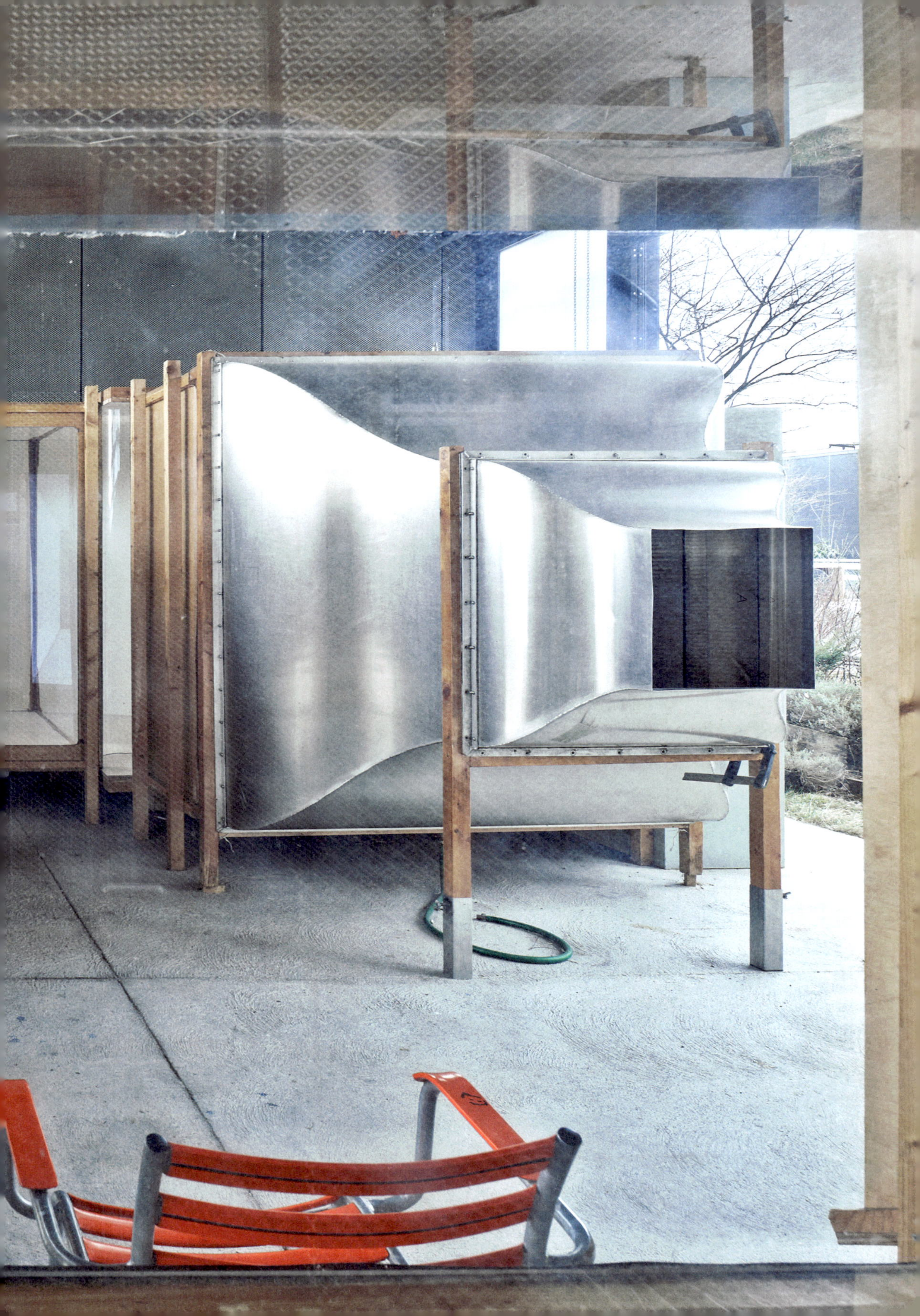

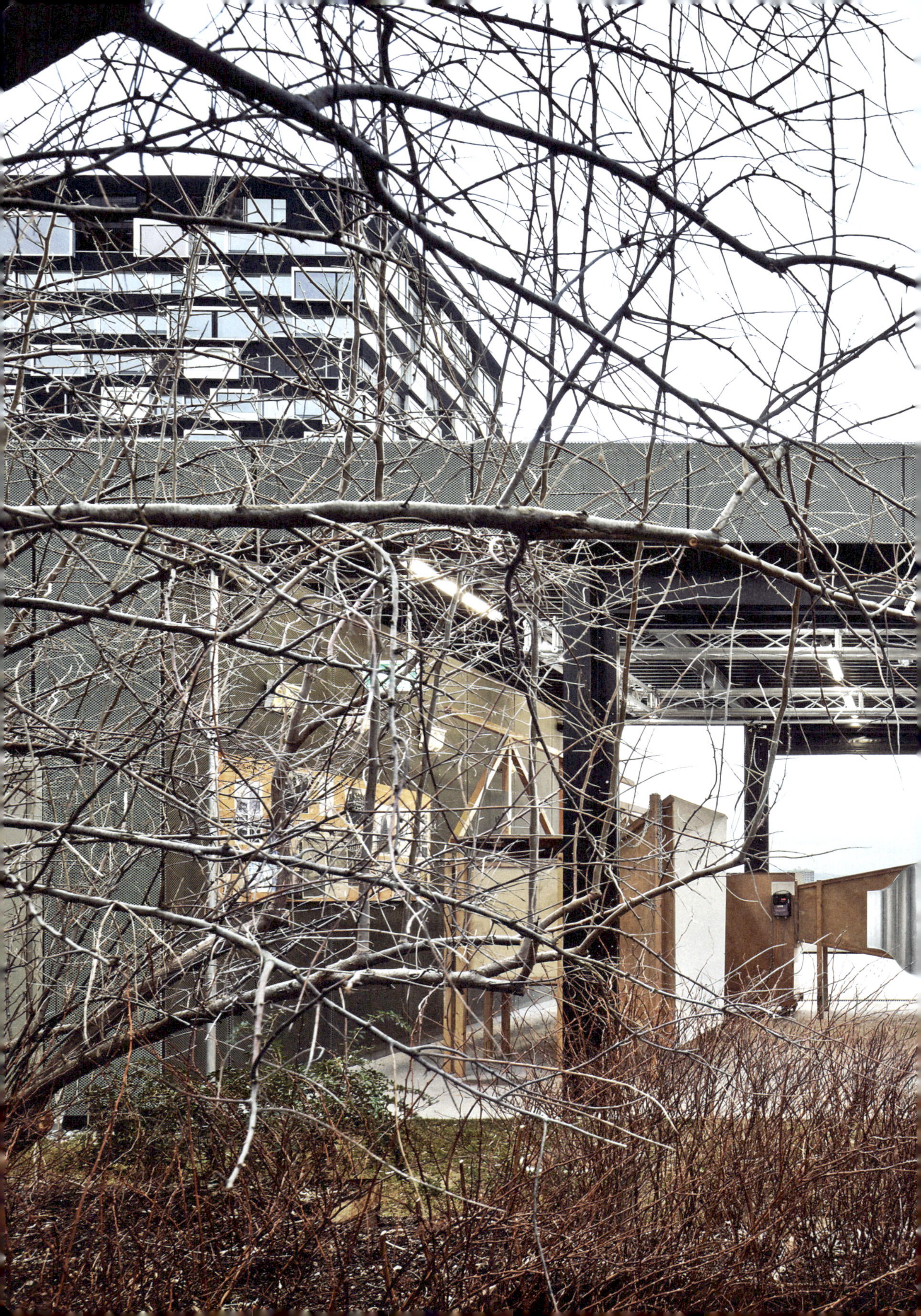

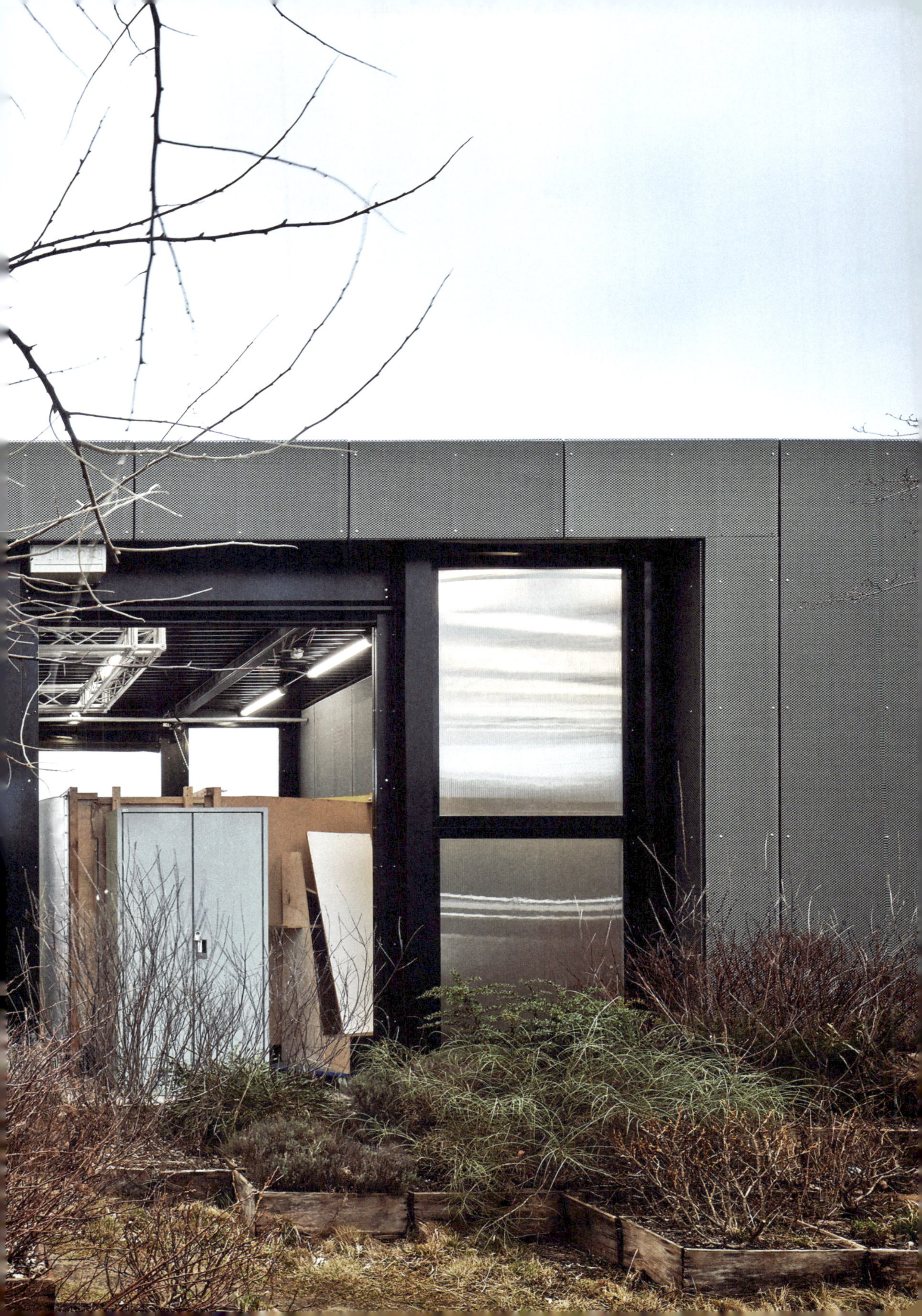

You do not love if love does not disappoint the expectations you saddle it with.
Marcus Steinweg, 2010 (original in German)

If facts destroy theory, then all the better for theory.
Viktor Shklovsky, 1927 (Trans. Alexandra Berlina)

In many places and for quite a few years now, the arts have been expected to do research. "Exactly", say some, "and we always have!" But then others cry, "God forbid, that's the downfall of art!" Neither of them is completely wrong, I would say. And, as always, it depends. So this book does not give a general answer to an overly general question, but presents a very specific attempt to make sense of the confrontation between art and research in an exemplary way. And it is a second attempt. My first one was articulated ten years ago under the title 'Kunst als Forschung. Ein Versuch, sich selbst eine Anleitung zu schreiben' (Art as Research. An Attempt to Draft Some Instructions for Myself, Dombois 2006) and experienced an astonishing reception career, was read a lot, translated, and reprinted several times. But the attempt has become unrecognisable to me. Instead, I see a hijacked discourse under the name 'artistic research' in which things become established that do not interest me as an artist or even beset my work. I am disappointed.

Okay, so let's start anew; let's go back to the beginning; let's forget what has been said so far again. The original question was: can one set up a force field between the arts and sciences that overstrains both sides in a meaningful and productive way? Can the presence of scientific disciplines entice me as an artist to take paths that are new to me? Can I open myself up to the inconceivable through a confrontation with the strange? The expression *art as research* is, as I understand it, a discomforting assertion. For both sides. And only when you can scarcely bear it are you right. Only when you no longer know in advance what you will say can the pledge of act-uality be honoured. When you articulate a meaningful posit into a force field, then actuality can become qualified. That is what I am interested in. And I believe that such movements are instructive for anyone who wants to forge ahead into unknown territory.

The Wind Tunnel Model is a suggestion for such a force field and a number of posits. The form it has taken since 2013 is influenced by the respective state of reflection, by the team I was able to gain for it, and by the conditions we are subject to in Zurich and Switzerland. And I believe that the specific aspect is not detrimental to the matter. Quite the contrary. So let's get right to the heart of the matter. Let's begin with the sections of this book, which are divided up according to the following different perspectives offered on the subject of models:

The introduction and the ensuing manual, which I wrote with Kaspar König, describe *The Wind Tunnel Model* in perhaps the most abstract form and with the most general claim. We hope there will be many imitators and completely agree with James Elkins, who said during a visit to our wind tunnel, "Every art university should have one!"

This is followed by a collection of five essays written by Jacqueline Burckhardt, Christoph Hoffmann, Dieter Mersch, Isabel Mundry and Jan Svenungsson, respectively. These five colleagues, who accompanied the project last year as a reflection group, examine the wind tunnel from the outside from their respective disciplinary perspective. The following eight essays are by the participants who in recent years have worked with the wind tunnel as a direct point of departure and who have literally stood in the wind. The authors are Haseeb Ahmed, Martin Burr, Julie Harboe, Kaspar König, Mirjam Steiner, Sarine Waltenspül, Reinhard Wendler and I. In between, I have included an essay about my team's mode of working for which I developed a model based on the man-engine (*Fahrkunst*).

Finally, the last section takes a documentary approach by listing the course of activities from January 2013 to December 2015 (the period when the project was sponsored by the Swiss National Science Foundation SNSF) and an excerpt from our list of publications. I have added an essay to this part, on the question of what advantages might result from research projects for the arts (and the sciences).

So readers can switch between a close and distant view of *The Wind Tunnel Model*. They can switch between the mode of the manual, the case study, or the overview. But they can also engage with individual positions within the framework and approach *The Wind Tunnel Model* in this way. Or to put it in other words: the research topic of our SNSF project is titled 'Size Matters – On Scale and Size of Models', and different modes of viewing that involve distancing, changes of perspective and other forms of scaling recur throughout this publication in different forms of distance to our work.

The form of the book reacts to this. It creates peripheral and intermediate areas in which readers have space for notes, supplements, corrections, new ideas. It is a workbook; it wants to be continued and transformed. We put things up for discussion and we want to be put up for discussion ourselves. Thought models have to be taken in the hand, have to be able to be screwed and filed. They are not set in stone.

The Wind Tunnel Model is the current form of my investigation of how artists can dwell in a research environment. It is manifested in a real wind tunnel that we built at the Zurich University of the Arts and that today is installed on the roof of the Toni-Areal building. But not just that – far from it. The specific location of the installation also plays a role, both topographically and metaphorically – for instance, the fact that we are installed on the roof of the institution, or that an escape route leads through our laboratory. The people who supervise it factor in, as well as those who visit, use, and reflect on it. The discourse that arises and is created around the artefact on the Toni-Areal roof also comes into play. And even the wind tunnel itself is actually already a first enclosure of the centre of the work, namely the wind that circulates in its interior and which the tunnel was built to guide and support. If we think of these factors that contribute meaning as the layers of an onion, then it is not easy to say where *The Wind Tunnel Model* begins and where it ends, both inwardly and outwardly. The horizon and the limits of the model are not easy to define, yet we are trying to put the model, insofar as possible, between two book covers.

But who can benefit from this?, some will surely ask. Why do we need a wind tunnel model, for Christ's sake? As an artist, I find questions about the benefits and utility of my work very unpleasant. But they accompany the research claim and so I don't want to evade them, even if I can't answer them, and surely not as a whole. Yes, perhaps there is not just one answer to this question? Perhaps we need an enormous number of answers, to wit, as many as there are people? This gives me hope, and so, esteemed reader, please allow me to answer them individually and very personally to you, at least for a selection of people:

For a citizen

Sit down in the measuring section of the wind tunnel. Now I'm going to turn on the motor and slowly increase the revolutions per minute. Then tell me under what wind conditions you can think best.

For an art historian

The first aerodynamic institutions and wind tunnels as they are known today were founded a short time after 1900, and the futurists enthusiastically appropriated figures from the aerodynamic visual language – e.g. the supersonic cone – as an expression of speed. Thus we find different aerodynamic signatures e.g. in the works of Russolo as well as those of other painters, a celebration of racing cars, aircraft and bullets. But does a wind tunnel always have to be operated at full speed? What are the images of slow wind like that the futurists apparently sorted out? What would a 'modern' visual language have been like that instead of getting off on the macho phantasies of faster, higher, further had rather concentrated on the slow, on the poetic, for example on cigarette smoke that rises or the flight of a snowflake? What kind of modernism would have resulted?

For a designer

With the first wind tunnels, a new era of form finding also began in design. Because for the first time in history 'good form' was not solely determined by taste or basic mathematical forms, but there was a machine that could measure the difference between a more streamlined and a more resistant form. This left behind traces in the design, which are especially visible in automobiles, but also in consumer goods, from the 1950s for example (cf. Lichtenstein 1993). Knowledge and use of this form-finding machine could therefore really be a component of every design workshop. What does the optimization of a form mean when an experimental set-up is integrated into this process that interrogates not only designers and their customers, but also a third authority – so-called nature?

For an aerodynamicist

Wind tunnels themselves also display a wealth of forms. Depending on the scaling of the test objects, the questions involved in the experiments, the speeds and so on, new tunnels are continually developed and built. The encounter with the artists' wind tunnel is a non-verbal form of fundamental questioning and reflection. Apart from the streamlined design, it examines the essences of the facility – what *is* a wind tunnel to begin with? And thus it leads right into the field of basic aerodynamic research, which is dwindling dramatically. What has become of it?

For a historian of science

Wind tunnels are the dinosaurs among natural-science laboratories. The Wright Brothers' wind tunnel at MIT, for example, has operated almost exactly the same way since it was completed in 1938. That was more than 75 years ago. Yes, and even one of the very first professional wind tunnels, Gustave Eiffel's *Soufflerie* in Paris, has operated uninterruptedly since 1912. Where else in the history of science have scientific laboratories been run for so long? Where else have the conditions of an experimental system remained constant for such a long time? Perhaps the different wind tunnels are suitable as twentieth-century index fossils, like ammonoids were from the lower Devonian to the end of the Cretaceous period?

For a science theorist

The Wind Tunnel Model always acts on several levels. The wind tunnel demands manual action, requires that one exposes oneself to the wind, and at the same time reveals that these actions are meaningful from an abstract perspective as well: thus, for example, in this way it becomes evident that the modus operandi of models not only lies in their being used as models, in other words artefacts, themselves, but also in the fact that their surroundings are prepared by the experimenter. The wind tunnel is a metaphor machine that reveals a lot about scientific experiments, the role of scientists and their recording devices (e.g. the camera), as well as the question of models and scaling. And as disturbing as it might be to intervene manually, here the sign as a trace of the concrete thought corresponds all too often with the concept as a trace of abstract thought. Thus a stay in the wind tunnel goes far beyond aerodynamics and could very well be a viable explanatory model in and of itself. What are models and what are model experiments? How does scaling work as a basic epistemic technique of rational cognition?

For an artist

I am fascinated by the wind as an artistic medium. Its invisibility and fleetingness appeal to me, as well as the challenge of giving a form to this nothingness. A formwork construction as if for plaster casting, the flow of the air that is modulated through the wind tunnel walls and is always moving. For the wind tunnel that appears before our eyes like a sculpture is always only the reverse side of the actual form, which is preserved inside the tunnel. What kind of 'second-order' sculptures are they?

And at the same time the question of practice presents itself – how can I work with several artists and authors at the same time on a joint work without either dominating with my aesthetic aspirations or sacrificing them? Does work on 'second-order' artworks make us more tolerant of other artistic designs?

For an artist researcher

In connection with the research claim of the arts I am interested in turning away from a claim to knowledge (is it really urgently necessary to justify the arts epistemically?) and towards research that serves artistic production. What must a research result be like that is not yet intended to be art and precisely for this reason motivates and supports other artists in their artistic practice? What would such a not-yet-art look like that also supports the artist on the art market, for example the way aerodynamic research at universities furnishes the energy sector with new ideas? Research by artists for artists? And how would art itself develop if art universities did not only foster individual talents but also artistic disciplines?

For a politician

And for those who prefer to calculate, the answer is: the art market is not a negligible part of the economy. A kind of research that supports production for this market and makes it competitive in an international comparison amortizes itself in the same way and via the same kind of yields and cash flows as technological developments do in spite of the fact that costs of research for technological developments are sometimes very high. So the formula is: we need more wind tunnels!

For we need to think more – rather than having thought. Research means to continue. And that is something that fascinates me about the natural sciences, that at the end of so many publications it says: "Here more research is needed." "It is worth thinking more about this matter." "Here we perhaps have to formulate better although we tried as best we could." The situation in physics is no different than with Wittgenstein, who believed he had solved all problems once and for all with the *Tractatus*, only to begin anew again. The so-called history of progress of the sciences has no end. And it has no definitive formulation of the knowledge it has gained. Research is thinking that questions.

We remain true to doubt. We distrust simple solutions. We distrust recipes that do not challenge users but make them act mindlessly. Tell me about things that are not clarified, about which it is really worth thinking and where I am challenged as an independent, free-thinking individual.

Of the ancient Greek temples, the *Heraion* in Olympia is my favourite. It is of the Doric order and one of the oldest temples whose ruins we can still visit today. However, it is also one of the youngest Greek temples. Originally built of wood, it was transferred from one material to another, from wood to stone, very slowly, piece by piece, stone by stone, over half a millennium – as late as the Roman imperial period, Pausanias was shown the last oak column. And the stone columns reflect the changing tastes at the time of their creation. Whenever a wooden column was replaced, the stone replacement was designed *à la mode*, with the result that the diameters now vary from 1.00 to 1.28 metres, and the number of flutes and the form of the capitals also vary. I admire the Olympians for their courage and their correct aspiration, in my eyes, of always articulating the best possible form at any time. And by carrying out research with precisely this aspiration I survive as an artist: research that is free enough to continue and at any time to look for and formulate the expression that is correct at the moment. And so I ask: where is the wind when it's not blowing?

When one writes and publishes a book, one establishes a relationship with many people. I would like to thank the authors for their texts and their open and friendly cooperation during the preparation. Special thanks go to my team for their trust and the unusually giving atmosphere in which we could work all these years. A special thanks to Isabel Mundry for her idea to form a reflection group. I thank Viola Zimmermann for designing the book and for working together with me to find the right form. I thank Sebastian Viebahn, who oversaw the translations. The series of digital renderings from the perspective of the interior of the wind tunnel was possible thanks to the initiative and the 3D scans of Jonathan Banz, to whom I am grateful. I thank Cornelia Mechler from the publisher for her excellent supervision. I thank the Swiss National Science Foundation, which accepted that our project was open-ended (our application only named a procedure, but did not define any specific results that had to be fulfilled) and which hopefully feels rewarded by the unusually high output of the project. I thank the Zurich University of the Arts and particularly my superior Christoph Weckerle for his trust in my work and my claim that the wind tunnel is important. And I would like to thank the housekeeping and other university service staff, to whom our wind tunnel always posed a big problem and challenge. And once again I would like to thank the Zurich University of the Arts for the generous financial support for the printing of this book.

English translation by Burke Barrett

References

Dombois, Florian, 2006, 'Kunst als Forschung: Ein Versuch, sich selbst eine Anleitung zu schreiben' in Hochschule der Künste Bern (ed.), *HKB | HEAB 2006*. Bern: Bern University of the Arts: 23–31.

Lichtenstein, Claude, and Franz Engler (eds.), 1993, *Stromlinienform, Streamline, Aérodynamisme, Aerodinamismo* (exh. cat.). Zurich: Müller.

... when I said I must go, and asked him to further me on my way, he made no sort of difficulty, but set about doing so at once. Moreover, he flayed me a prime ox-hide to hold the ways of the roaring winds, which he shut up in the hide as in a sack – for Jove had made him captain over the winds, and he could stir or still each one of them according to his own pleasure. He put the sack in the ship and bound the mouth so tightly with a silver thread that not even a breath of a side-wind could blow from any quarter. The West wind which was fair for us did he alone let blow as it chose...

Homer: Odyssey X, 17–26
(Transl. Samuel Butler)

The following text was written in two parts: Florian Dombois wrote the first four recommendations, Kaspar König the second four. Moreover, the text operates on two levels: it is devoted to the actual structural design of the wind tunnel and to the question of how collective, poetic laboratories can be created. This yields a *matrix*. And *matrix* (lat.) denotes a breeding female.

1. Put a void in the centre so that all users can project themselves and their world onto it.

As artists, we are interested in poetic space, the space in which something can arise. Poetic space is not a workshop, not a studio. It is not functional, i.e. geared to something specific, but, if anything, is directed to a certain *je ne sais quoi*. Poetic spaces have a tough time in a world in which you have to say what you mean. They need protection from false intrusiveness. One possibility is to surround them with a tunnel. For wind tunnels are ambivalent objects in the best sense: they promise to give an answer to those who bring something with them and thus subject themselves to the wind tunnels. However, the ones who ask the question always have to make a twofold investment: they decide on the model that is to be tested, and they decide on the means to be used to visualise the wind. Thus the people who ask the questions make themselves visible to those surrounding them in several respects. Acting in this way, they show themselves and their discipline.

The Wind Tunnel Model focuses on the encounter between artistic and scientific disciplines. Up until now, there have been two main locations for such encounters: the lecture hall where the sciences are presented, and the exhibition space, stage, or concert hall where the arts are presented. These places are asymmetrical; they favour one discipline over another. And they shift the performance to a different reception context: artists are perceived differently in a lecture hall; and so are scientists in art spaces. How can a balance be struck between the roles? Isn't poetic space dependent on the ambiguity between the explicitly visible and the simultaneous promise of metaphorical meaning coupled with it?

2. Define a theme that is so multilayered that it can dwell in as many intellectual spaces as possible.

We examine models and their effects in different contexts. This is an endless issue because models can be found everywhere. So we concentrate on one aspect: scaling. This very old yet still topical technique of cognition has a striking correspondence in the wind tunnel, so-called scale effects. As György Kepes explained in the 1960s, when "aeroplanes are designed [...] they make small models, not in the exact dimensions that they will be, but they have to distort it to perform the small model scale the same as it will be in the large scale" (Kepes s.a.). And we take this correspondence even further: for us, the movement between the concrete situation at the location and the simultaneous metaphorical meaning in the mind is itself a scaling movement. *Mens et manus*, handcraft and intellectual work acting together to achieve productive scale confusion. As scalers, we always dwell in the small and the big at the same time; our thoughts vacillate between the concrete and the abstract, between the explicit and the metaphorical. And that is a good thing, because in our context it is necessary to be specific and non-specific at the same time – only then can something truly unexpected arise, can something new spring to mind by chance. We need to devote time to this.

3. Form a work group that contributes and builds from the different disciplines and perspectives of its members.

People are needed who turn the artefact into a thing of reality. For just as the walls of the tunnel direct and shape the wind, people shape discourse. In our constellation, the core group consisted of three artists and three art and literary scholars with different degrees of experience. In the composition of the group, ensure there is a balanced mix of disciplines and characters. The force field of the individuals determines many of the things that can occur later. They form the plinth of the sculpture with their discourse.

Moreover, find a good rhythm of cooperation. We met once a week to discuss technical matters and once a month to discuss content. In between we went our separate ways. In tune and off tune – that is the beat of such cooperative work. Also discuss the rules of quoting at the beginning. Which results belong to whom and how can one honour the contributions of the others? Here the variety of disciplines will help you. May each one use the stream of the tunnel to articulate the specific form of his or her discipline, but never forget to express gratitude for the implicit and explicit contributions of the others. Only this ensures cohesion on a sustainable basis.

4. Always think about the wind inside when you give the outside a form.

A wind tunnel resembles a sculpture. In fact, however, it is a support structure, a scaffold, a formwork determined by the invisible interior. Take this invisibility seriously and concentrate on this empty space. Then your hands will take the right aesthetic decisions. Do not build based primarily on a blueprint, but build for the wind. Then you can build together with several people at the same time. Then you don't need an architect to provide the layout and to monitor the others based on it. Although it is good to have a protagonist, a foreman, think of it as a kind of Gothic builders' hut.

I have never seen a wind tunnel I would not have liked if it were a sculpture. Indeed, I like wind tunnels better than the streamlined products that have been developed inside of them for around a hundred years. Wind tunnels are the flipside of a modern age geared to optimization. They participate in creating in the 'beautiful efficient form', turn their smooth side to the model and at the same time show us their backs on the outside. Just like that. And they are attractive in an erotic way. Not only the round forms. They suck and blow. And it is hard to decide whether to give the wind tunnel a male or female name. The one can't be had without the other.

1. Clarify your relationship to existing wind tunnels.

The engineer Henrich Focke, famous for his contributions to the development of the helicopter, built his own wind tunnel in 1960, after his retirement, in a backyard shed in Bremen. Financially poor, he used cheap post-war materials such as hardboard, cardboard tubes, wires and so on. Many things had to be improvised, e.g. a choke for a drill made by connecting in addition a series of light bulbs, or a measuring device consisting of baby scales, self-cast weights consisting of lead from a marksmen festival ground and cups filled with oil for cushioning. In October 2012, the two of us visited his now lovingly restored wind tunnel and learned that the facility is nevertheless extremely effective. For instance, his life-long aerodynamic studies enabled him to give the propeller, which he filed himself at his kitchen table, a form that is more efficient than all of today's mathematically calculated models. Form follows function.

This visit was really uplifting for us. We decided to build our own wind tunnel as a modular and mobile one so that it could be moved easily. The width of the exit door at our wind tunnel's first location in the Hafnerstrasse in Zurich became our gauge. A maximum of two people had to be able to carry each component. We gathered many images of wind tunnels and found countless styles, ranging from cold and technical to ornamental and playful. It seemed to us that each wind tunnel was last but not least a caricature of the previous one. We did not find a standard form to which all of them were oriented, but innumerable variations on the theme.

2. Organise a motor.

At the beginning of our wind tunnel, there was a fan. It came from the former Progymnasium in Bern (today the artists' house PROGR), a secondary school where Albert Einstein once conducted experiments.

Its roof was refurbished and the air-conditioning renewed. A radial fan, weighing around 70 kg, was flown out by helicopter. This fan was our first component. And the first thing we used it for was to kindle a fire in our courtyard on Hafnerstrasse. And our curiosity.

In parallel, I drew our wind tunnel (*Wind Tunnel Bulletin* 2013: 8–11) and modelled it prior to construction using Rhino software. All of the components, including the motor, are conceived as modules: the settling chamber, the straightener, the nozzle, the test section, and the return channel.

3. Formulate the requirements you want to place on the wind tunnel.

What aesthetic demands do you want to place on a wind tunnel? Can you give the machine a 'wind expression'? To what extent should it be recognisable as a wind tunnel? Is there something like a typical wind tunnel? When they see our tunnel, many people ask "What is that"? although it is one of the most typical wind tunnels I know. Is our tunnel an iconic wind tunnel?

To be sure, our wind tunnel is a laboratory at a university of art. It has to run like clockwork; it has to provide wind for our research group, as well as for students and faculty. Space is needed for this. And actually also calm air all around...

We want to see a lot of invisible wind. So our tunnel is transparent in several places. The materials we used were Plexiglas, plywood, and spruce strips. That way, we can screw on more things later. Our laboratory is covered, but outdoors. So we coated all of the wooden parts with spar varnish. The advantage of this is that you can experiment with liquids in the wind tunnel without any problems. And it makes the surfaces smooth. In the bend I installed vertical guiding vanes to guide the wind around the corner better, i.e. with less turbulence. In front of the nozzle we have a straightener that calms the turbulences that remain before the wind is compressed in the nozzle and thus accelerated. For calculating the straightener (and for other construction issues) I can recommend the book *Low Speed Wind Tunnel Testing* (Barlow et al. 1999). For the shape of the nozzle it is easiest to use the 7th order polynomial curve, which you can adapt to your own dimensions (Carbonaro 2013: 7).

4. Develop a store of visualisation and audification techniques.

Wind only shows itself when other things cross its path. There are classical visualisation means such as sand and smoke; but feathers, soap bubbles and polystyrene beads also show the wind well. And even snow is an excellent wind visualizer. An example is strong snowfall spraying around a building.

Furthermore, so-called tell tales made of wool or silk are good indicators. Ideally, many of these should be used at the same time. They should be put next to each other or in a grid, so that larger movements can be observed. Light is important! If you work with smoke, for example, you shouldn't illuminate all of it, but only cut through the cloud with a sheet of light. Less is more. This can be done using a laser, or even with a projector and a slide into which two razorblades are glued which leave a precise gap.

For us, the acoustics were also very important. Which is why we have a quiet motor. We can hear the wind with our ears, as well as with microphones. And we build wind instruments. With them, you can also see invisible wind with your eyes closed. For all of our experiments are always also thought experiments. Professor Ming told us this story from China:

> Two monks see a flag in the wind. The first monk says: "The flag is flying." "No", says the second, "it is the wind which is flying." A third young monk hears it and says: "Excuse me, the mind is flying."

And we don't only work on streamlined thought, but also time and time again on making it turbulent. The uncomfortable is the order of the day. We have occasionally called this *designated discomfort areas*.

English translation by Burke Barrett

References

Barlow, Jewel B., William H. Rae and Alan Pope (eds.), 1999, *Low Speed Wind Tunnel Testing*, 3rd ed. New York, NY: Wiley.

Carbonaro, M., 2013, *Design and Construction of a Small, Cheap, Portable Wind Tunnel for School Labs*, Technical Memorandum 55, Von Karman Institute (rev. version February 2013).

Kepes, György, s.a., film recording *Kepes I No. 91*, Kepes Archive of the ACT programme at MIT, Boston, Mass.

Wind Tunnel Bulletin #1. 2013 (November). Zurich. (Online at http://windtunnelbulletin.zhdk.ch/01/).

Let's not deceive ourselves: "This is neither a real wind tunnel nor a real art work", says Florian Dombois, head of the research group 'Size Matters' at Zurich University of the Arts.[1] But then what is that thing on the roof of the university building? What is it *in reality*, and is there anything it is for *real* at all? The members of the Reflection Group devoted a considerable amount of time and energy to these questions in the joint discussions. One could almost believe it is an intentional puzzle aimed at confronting the participants with their need to categorize. In this case, the thing on the roof would be incentive for a philosophical experiment: how long would it take before a dozen people stopped looking for a clear order? At what point is the *furor systematicus* exhausted? What is not supposed to be a real wind tunnel nor a real artwork would thus prove to be a stumbling block.

My thoughts about it are simpler. What the thing on the roof is about can initially be answered by listing the materials used: multiplex and fibreboard panels, square timbers, aluminium sheeting, adjustable metal legs, hard plastic, Plexiglas, screws, glue, self-adhesive paper tape, spar varnish, a motor, a control panel. One can even be simpler than that. Those who don't know anything about it initially notice from a few metres away a box of substantial size that is closed on all sides and propped up on a trestlework. Moving closer, I observe that the box winds around four corners and ends in a nozzle-like opening. Across from the latter, around one metre away, is a second opening with a considerably larger cross-section. Walking through the gap, I enter a completely closed-off inner area. From there, I can clearly see how the walls of the box sometimes get wider, sometimes narrower. A control panel is attached to one of the walls. When the rotary switch is moved to the right, a motor starts running making a fair amount of noise.

All in all, it seems to be a technical facility, but what it is for remains open. On two sections, windows extending from one side to the other are integrated in the hull, but there is nothing you might observe. Nor is there anything to be seen in the space between the two openings. Even when you come closer, everything

1 All of the remarks of the participants and other players are, if not stated otherwise, taken from the records of the 'Size Matters' Reflection Group in the course of 2015. Separate evidence is not supplied hereinafter.

seems unchanged. Only when I hold my hand in the space do I feel a draught. This much can be said: the thing on the roof is a rather elaborate blower, like an oversized hairdryer. The further the switch is turned, the stronger the current that comes out of the nozzle. A few things are lying around: the wind gauge indicates 10 m/s. Coal dust is distributed on a tray and takes on a fixed pattern in the airstream. Smoke is blown into one of the openings and after a while comes out of the other. One catches sight of a slip of paper on which the following has been pre-printed: “Experiment”, “Author”, “Date”, “Place”.

The question of why the box is not a *real artwork* can now be answered readily: the facility is not sufficient for itself, does not stand for itself; it serves a purpose, it *works*. During construction the builders wanted to ensure that one can do more with this thing than just contemplate at it. But why isn’t it a *real wind tunnel*? For as long as it serves its purpose, it doesn’t matter whether the facility is standing in an aerodynamics institute or on the roof of an art school. One reason might be that while the construction works, it does not work quite like *a real wind tunnel*. Something about it makes it different from the wind tunnels that were inspected in the run-up to the project. Haseeb Ahmed, a research assistant for ‘Size Matters’, drops a hint in that direction when he says: “Maybe it is a bit like a reinvention of the wind tunnel. Like drawing an object that you only know from memory and that is also what makes it interesting.”

With this remark in mind, one can imagine the thing on the roof as a washed-out copy of a wind tunnel. The facility brings together, like a composite image, all of the impressions gained before the work began during visits to other wind tunnels, the many photographs and sketches taken from literature on the subject, and finally the numerous stories connected with it. So the box is not simply a *wind tunnel*. The box is the tunnel of tunnels, the average wind tunnel resulting from all the others. Perhaps the historian of science Hans-Jörg Rheinberger had this in mind when, upon being confronted with the box, he remarked: “The wind tunnel [of the ‘Size Matters’ group] is a model of a second order and at the same time unsettlingly specific.”[2] Viewed from this perspective, one can say that the facility is not *a real wind tunnel* because it brings into view and makes perceptible that which constitutes a wind tunnel. This facility works – and has to work – and it has to do so precisely because it is not to be operated like a normal wind tunnel, but to be taken into consideration as a particular facility.

Could we go so far as to say that the large wound box is a wind tunnel in its pure state? The outer appearance speaks against

2 See for the quotation the blog on the project website: http://blog.zhdk.ch/windkanal/tag/hans-jorg-rheinberger/ (accessed 19 March 2016).

this interpretation. Nothing is polished and perfectly packaged. Screw heads peep out; there are traces of glue here and there; the square timbers of the frame visibly protrude; the facility consists of segments whose joints can be recognized everywhere. The whole thing seems very provisional; one thinks of a backyard shop or a backstage area rather than a laboratory or research institute. Nearly all of the materials can be bought at a DIY store, and the only thing that really glistens is the customised nozzle made of aluminium sheet metal. If this is a wind tunnel in a pure state, then in this case pure means raw or naked. Again: "Maybe it is a bit like a reinvention of the wind tunnel. Like drawing an object that you only know from memory"; and while I read this sentence, I think of descriptions of the first wind tunnels, of names like Francis Herbert Wenham, Horatio Philipps, Ludwig Mach, Hiram Maxim, the Wright brothers in their bicycle shop in Dayton, and a few others, always the same picture: wooden boxes, glass, metal nozzles, simple fans, simple aids, scales, smoke, heated air (Randers-Pehrson 1935).

All of this comes to life again on the roof of the Zurich University of the Arts. It is more than a *reinvention*; it is a *re-enactment*. Or a kind of phenomenological reduction. This doesn't occur in words, however, but materially. The *black box* wind tunnel is stripped; the components are openly exposed; there are flow straighteners, baffles; if you want, you can stick your head into the orifice, and with one handgrip you arrive at the fan motor; when you turn the switch to the right it starts running; there is a pack of smoke-generating agent on a stool; chains with strips of cloth hang from the ceiling. But what can you find out in this way? What does a wind tunnel consist of essentially? The reference to the purpose doesn't help us make any progress. It is true that with the help of a wind tunnel you can determine the air resistance of a body and its optimum aerodynamic form. It is also true that in this way you can examine processes in the air around a body, flow lines and turbulences, compressions and rarefactions. But then you move away from the wind tunnel. The latter as a whole is not considered, but rather you look at the measuring distance and analyse the recorded instrument data, the photos and videos.

What is a wind tunnel? I propose a detour. "There is too much wind up there." Jan Svenungsson, member of the Reflection Group, artist, and model airplane builder, cuts to the chase. The spot on top of the university is extremely ill-chosen for a wind tunnel. The covered area, about ten metres wide, eight metres deep and four metres high, is open on two sides. On the right and left it is framed by functional constructions, which like a funnel catch every draught that blows over the ground, and guide it through the opening. "We want to have the area protected from the wind, but we are still waiting for permission from the school to do so."

With his apology, Florian Dombois reveals a paradox: to *really* be able to operate a wind tunnel, you have to protect it from the wind. In other words: a wind tunnel installed outdoors is an absurdity. It needs four walls and roof; otherwise the effort is wasted (I think of the shed in which Ludwig Prandtl began his work in Göttingen in 1908; and I think of the luxury building that Gustave Eiffel moved into in Auteuil soon afterwards).

What is a wind tunnel? Like every laboratory, it is an installation for producing extremely artificial things. The wind of the wind tunnel has little in common with the wind that blows over the roof. As Florian Dombois says: "The straight wind of the tunnel does not exist in nature." To be more precise, a wind tunnel does not produce wind, but an airstream that is modulated by baffles and flow straighteners and instrumentalized. This airstream is not artificial simply because it occurs outdoors only by chance, if at all. It is artificial because in the wind tunnel it is available as otherwise perhaps only for a sailing boat or the rotors of a wind turbine (cf. Heidegger 1977: 716). Thus two things are significant about a wind tunnel. Something is provided in it that otherwise is hard to control. This can only be done by producing a second wind, a new wind; a predictable, uniform and hence improbable current.

The installation on the roof – "[t]here is too much wind up there" – shifts the airstream into the realm of the questionable. The wind, which is disturbing, gives the wind in the wind tunnel a specific presence. The airstream no longer only provides its service, but also gains peculiarity as an airstream. For what wind can be turned on and off? What wind blows so controllably? But this does not mean that the air current in the tunnel has nothing to do with the wind that sweeps over the roof. No one would build wind tunnels if experiences with models could not be transferred to the conditions outside of the tunnel with sufficient reliability. Rather, since the airstream in the tunnel proves to be a special wind, an everyday phenomenon, *the wind*, like that currently wafting through the trees outside of my window, conversely becomes an object of observation. The airstream constrained in the wind tunnel calls our attention to the wind outside (we notice it as a disturbance on the roof) and at the same time becomes a special representative of that which we generally undifferentiatedly term *wind*.

The engineer of 'Size Matters', the designer and musician Kaspar König, is not interested in whether the facility is a real artwork or a real wind tunnel: "Our question is, 'what is the wind?'" We put a piano askew in the space between the two openings, or to be precise, the rest of a piano that only consists of the back wall with the string body. When we turn the switch to the right, the motor hums, the paper tape with which the segments of the wind tunnel are sealed off gives rise to a characteristic noise; close to the piano we hear a soft sound. "The wind is the actor and we want to

References

Randers-Pehrson, Nils Henrik, 1935, *Pioneer Wind Tunnels*. Washington, DC: Smithsonian Institution.

Hacking, Ian, 2008, *Representing and Intervening: Introductory Topics in the Philosophy of Natural Science* (1983), 22nd edition. Cambridge: Cambridge University Press.

Heidegger, Martin, 1977, 'The Question Concerning Technology (1955)' in *The Question Concerning Technology and Other Essays* (1953) (transl. by William Lovitt). New York: Garland Publishing: 3–35.

see what it does." König turns things on their head. He unleashes the wind, not from its special constitution as an airstream directed in the tunnel, but from its inconspicuous givenness. It is only a small shift: He aims not at producing the sound, but rather at having the sound characterize the airstream, the wind, by opening up an acoustic approach to it. Does this make sense? A thought of the Canadian philosopher Ian Hacking comes to mind. In *Representing and Intervening* he writes: "The physicist George Darwin used to say that every once in a while one should do a completely crazy experiment, like blowing the trumpet to the tulips every morning for a month. Probably nothing will happen, but if something did happen, that would be a stupendous discovery" (Hacking 2008: 154). I suppose that if Hacking heard about the wind tunnel on the roof of the ZHdK, it would be a pleasure for him to watch the goings-on there.

English translation by Burke Barrett

I'm on a vast, grass-covered field. A light breeze in my back. I hold the soft handle of a 50-metre towline[1] in my hand and tense it. My friend raises the model airplane towards the wind. I give a signal, turn and start running against the wind. He lets the model go and it quickly rises at a 45-degree angle. I slow down as soon as I feel the right level of tension in the line. When the plane has reached a point straight above my head I hold it there, gauging its behaviour through the line while looking at it and our surroundings. Any birds upwind? Circling? Other starters? Obstacles? Anything going on?

I'm hunting thermal. Rising air.

I decide to make a circle. I pull the plane over my head in front of me and with a small jerk I relieve the tension in the line, at which point the model's spring-loaded starting hook (onto which the towline's 15-mm diameter metal ring is hitched) snaps back and sends an impulse to the model's microcomputer, the 'timer', which steers the servos that operate the plane's control surfaces during start. Servo 1 decreases the incidence value of the horizontal stabilizer slightly from 11.5 to 10.4,[2] causing the plane to raise its nose imperceptibly; servo 2 moves the rudder at the back of the fin from its straight tow position of 21.0 to 48.0,[3] which causes a sharp right turn; servo 3 changes the incidence of the right half of the wing from 24.0 to 66.0,[4] aiding the model's pivot by making the right half of the wing act as a brake as well as preventing a spiralling motion towards the ground.

The model turns sharply to the right. I hold the line slack. Depending on the atmospheric conditions of the given moment, combined with how I handle the towline (pulling, holding it loose, jerking in different ways) the circle will be tighter or wider. I keep my eyes on the model, constantly analyzing its behaviour: is there steady air? Is it rising or sinking? As the model completes its circle and turns towards me, I stretch the line taut again: the hook moves back to its former position. The servos return the control surfaces to straight tow settings. I may be standing still, simply

1 Made of polyethylene fibre and 0.65 mm thick.
2 The incidence and rudder values quoted do not correspond to degrees, but represent much tinier changes – not proportionate between the servos. In this case the angular difference between stabilizer and wing is increased by about 0.2 degrees.
3 Here the rudder moves from a straight position to about ten degrees right.
4 Which means the incidence of the right wing is now about one degree higher than that of the left wing.

using the wind's resistance to bring the model back to a good position, or I may have to move to regain control. I may also want to move in order to reach a better position on the field. Where is the air that I need?

The manoeuvres described will usually be repeated a number of times before I feel confident that I have detected a thermal – that is: a bubble of air slightly warmer than its surroundings, which has begun to rise due to its lower density. This is where I want to launch my free flying glider model. This is where it belongs. Rising. Soaring.

When I have convinced myself I have good air, the launch needs to be preceded by one last circle during which I will 'pull' the model round at a lower altitude and on a taut towline. Moments after the plane has started to turn towards me, at about 7 o'clock (not yet in a straight position) at a height above ground of perhaps 20 metres, the hook now in its forward position, I start running as fast as I can – causing the model to accelerate rapidly. The tension in the towline rises immediately and at a load of about 7–8 kg, the timer instructs the servos to move control surfaces to 'zoom position', which puts the model in a modified straight tow attitude with maximised speed settings and a slight rightward tendency (servo 1 increases stabilizer incidence to 20.3; servo 2 puts rudder at 29.3 and servo 3 reduces right wing incidence to 23.0). The model, still on the line, is now speeding upwards in a centrifugal movement, thanks to me sprinting as fast as I can. As it reaches the sweet spot right above my head – the tension in the line now reaching 30 kg, its wings bent violently upwards, its speed around 40 m/sec – my body is fully engaged in the action. At exactly the right moment (I hope!), I open my right hand and let go of the line.

The complete drop in tension is registered by a pendulum mechanism in the hook, which immediately opens. The towline's metal ring falls off. The plane is freed from the line. The timer registers release and sets a quick series of pre-programmed commands in action. Immediately all three servos make radical changes: servo 1 moves the stabilizer to incidence value 00.0, which makes the back of the model drop and pitches the front violently upwards; servo 2 moves the rudder left, to 12.0, which acts to stop the centrifugal motion of the zoom action; servo 3 reduces the right wing incidence to 21.0. Given its speed and the stabilizer's setting, the model enters the first stage of a looping. However, after a little bit more than a tenth of a second – 00.12 seconds – servo 1 reverses the action of the stabilizer, moving it to incidence 33.4 instead, which in normal flight would cause a dive but which here acts to interrupt the looping action just as the model has reached a vertical position, which sets it racing straight upwards, until at 01.46 seconds – the model still pitched vertically but rapidly losing speed – the stabilizer is moved to 82.0, and an

extreme diving manoeuvre occurs. The model now performs a 'bunt':[5] it at once dips its nose, though more slowly as it has consumed its excess speed. It thus transitions to a horizontal position. When at 02.70 seconds the stabilizer is moved to the optimal gliding value of 24.5, normal glide flight begins and servo 1 will not move again unless or until the flight is aborted. Meanwhile, at 02.80 seconds, servo 2 moves the rudder to its gliding position, 27.0, causing the plane to make gentle curves; and at 02.82, servo 3 moves the right wing to 43.0, where it will stay. If all has gone well, the glider model has been catapulted to begin its flight at an altitude of about 100 metres, from its release at 50 metres. In the first 02.70 seconds after release, the model will have gained 50 metres of height or even more, before transitioning to glide flight in gentle circles at a speed of approximately 4.5 m/sec.

I'm in a hard-to-define space on the roof of an art university. I'm freezing: it's a winter's day. The room is lacking half a wall, it can't be closed and is open to the elements. I'm watching a wind tunnel being demonstrated.

In the rooms and halls and corridors below us, artists and teachers and students use electronic cards to access their studios and lecture rooms. All the rooms have windows to the endless corridors. Transparency is key. The environment is safe and efficient. The only exception to this model for a rationalized study space seems to be here on the roof, where this *singularity* has appeared: a wind tunnel surrounded by real wind often stronger than the artificial wind inside.

Central to the mastery of heavier-than-air human flight was understanding the role and science of airfoils. Each airplane wing cross-section consists of a curved shape designed to give the most favourable ratio of lift to drag in flight. This shape is called 'airfoil'. In the 1860s, Francis Herbert Wenham had understood, based on his study of bird flight, that "the camber of a wing should not necessarily be uniform, an arc of a circle, but should be thicker at the front and trail off at the rear..." (Goldstone 2015: 52). In 1871 he built the first closed-system wind tunnel to further his studies. Yet it would still be decades before a more general understanding of aerodynamic principles replaced various misconceptions.

After Wilbur and Orville Wright's disappointing experiences with their first glider at Kitty Hawk in 1901, they decided to build their own closed wind tunnel – likely after having read Wenham's papers. Thanks to Wilbur's genius for pragmatic conceptualization and Orville's outstanding craftsmanship, their 2-metre-long wind tunnel

5 The word 'bunt' originally denotes the "act of flying an aircraft in part of an outside loop" (Oxford Dictionary of English).

was "the most sophisticated ever built" (Goldstone 2015: 53). They used it to test more than 200 airfoil shapes and configurations and concluded that the much-respected research of Otto Lilienthal, on which they had based their earlier designs, was full of errors. This realization allowed them to move forward and achieve their dream of powered flight two years later: "As famous as we became for our 'Flyer' and its system of control, it all would never have happened if we had not developed our own wind tunnel and derived our own correct aerodynamic data" (Orville Wright, quoted in Goldstone 2015: 54).

Entering its glide phase at around 100 metres, ideally my model will continue to gain height in the thermal, leisurely spiralling upwards. If it is not gaining height, at least I hope it will find some neutral air, where it will be able to consume its impressive starting height to make its glide flight endure as long as needed. What really matters is to not end up in sink. If I happen to launch in bad air… the flight might be very short indeed.

There is no such thing as perfectly still air. Air is always moving. When a thermal lifts, there will be sink nearby. Learning to be good at flying my model is all about developing my sensitivity to experiencing and judging the present moment. I have to be acutely mindful of the here and now. I must strive to become an ever more sophisticated reader of the ether. I have to sense the minutest changes in the atmosphere. My goal is to make my model soar. It means a lot to me – but not to any other.

In addition to their astute development of aerodynamic theory, the Wright brothers' success depended significantly on their overall talent for *making*. Their pragmatic implementation of existing and development of new technology, through a process of trial and error, made possible the realization of what had until then been possible only in dreams.

In the visual arts at roughly the same time, the opposite tendency began to germinate: to disassociate the necessity of making from the fulfilment of the idea (the dream, the imagination). With his ready-mades, Marcel Duchamp began to use *naming* as a method, instead of making, in order to realize his work. In a retrospective interview with James Johnson Sweeney in 1946, Duchamp relates his activities of 30 years earlier and puts it this way: "I wanted to get away from the physical aspect of painting (…) I was interested in ideas – not merely in visual products. I wanted to put painting once again at the service of the mind" (Sweeney 1946: 19–21).

I first discovered the joy of making models soar as a youth, before I became obsessed by art and art making and gave

everything else up for that promise. Before this, for a few years, I was completely focused on building and flying F1A[6] free flying glider models, constantly taking part in local competitions. It would take more than thirty years before I came by a flying event again. Apart from one momentous decision, the basic framework and rules had not been changed. Yet the introduction of new construction methods and electronic timers had completely revolutionized performance and starting techniques. Before, modest forms of catapult launches had been attempted, which could result in gains of – at the most – a few metres. Wings built of wood (balsa and pine) could withstand a maximum load during launch of 3–4 kg, which severely limited the potential for catapult acceleration.[7] Built with carbon-fibre technique – still incorporating some balsa – the wing can withstand more than 30 kg. Spanning more than two metres, it still weighs only 200 g.[8] And in the last decade, launch performance has been enhanced to an almost absurd point, thanks to the introduction and implementation of a new type of airfoil design (for free flight models), known as LDA.[9]

Boris Groys wrote in 2005:

> Art today is defined by an identification between creation and selection. At least since Marcel Duchamp, it has been the case that selecting an artwork is the same as creating an artwork. That, of course, does not mean that all art since then has become readymade art. It does mean, however, that the creative act has become the act of selecting. Since that time, *producing* an object is no longer sufficient for its producer to be considered an artist. One must also *select* the object one has made oneself and declare it an artwork. Accordingly, since Duchamp, there is no longer any difference between an object one produces oneself and one produced by someone else – both have to be selected in order to be considered artworks. Today, an author is someone who selects, who *authorizes*. The author has become a curator. The artist is primarily the curator of him- or herself, because he or she selects him- or herself. And he or she also selects others: other objects, other artists.
> (Groys 2005: 93–94)

6 According to rules established by the FAI (Fédération Aéronautique Internationale, founded in 1905, two years after the Wrights' first flight), F1A free flight glider models must have a combined wing and stabilizer surface area of 32–34 dm² and may not weigh less than 410 g.

7 Also because the wings would flutter under pressure. Given the precision of the servo movements described in the first paragraphs, the wings must be extraordinarily stiff as regards torsion for the model not to veer off course and crash during launch.

8 And the stabilizer 7 g.

9 LDA is an abbreviation used for Low Drag Airfoil. Airfoils of this type are less cambered than traditional model glider airfoils, they have a 'fatter' nose and promote sustained laminar flow over the full chord of the wing, whereas a traditional, high-camber F1A airfoil causes some turbulence at the front in order to create lift. An LDA airfoil thus allows a much higher launch speed for the glider, while retaining similar performance to traditional designs during low speed glide flight. LDA models began to appear around 2005, but it took several years of trial and error before flyers gained sufficient knowledge on how to control them and avoid high-speed dive crashes.

To me, it seems clear that Marcel Duchamp created his readymades much more in the spirit of humorous play than as serious attempts to change art discourse – or even to make art at all. For us inhabitants of the information society, Duchamp's self-curation and constant play on naming and defining have become the most influential model of making art. It can seem as if there is no longer any need to actually make things yourself.

Sometime in the 1980s, the rules for free flight aeromodelling were changed, releasing the competitor from the obligation to build his or her own model. Combined with the widespread adoption of high-tech building methods and electronic programming, this rule change has led most ambitious flyers of today to order their models from a small group of highly specialized flyer-makers in Russia, Ukraine and the Baltic states.

At the same time, in the visual arts, the practice of commissioning the actual production of art works from fabricants (or assistants) near or far has gone from a sometimes necessary evil to a badge of honour for ambitious artists. They are eager to prove their unlimited ability to scale up production in answer to the global market demand they hope for.

For the practitioners of F1A glider model flying, the removal of the obligation to 'make' has accentuated the emphasis on conceptualizing (i.e. programming and adjusting) and personal performance (physical as well as intellectual). The need to make sense of the atmosphere, however, remains the same. At the same time a technological arms race has occurred with ever more complex and expensive models, which has caused a dramatic decrease in the number of practitioners. Only experts and fanatics remain. They, on the other hand, feel rewarded by performance hitherto thought impossible in the real world.

In visual art, the adoption of unlimited scalability of production has led to a shift in emphasis at the stage of reception, from judging direct traces of the artist's hand to admiring his or her ability to conjure up concepts and maintain production control. Whether looking at artists primarily focused on selling large numbers of products, or those presenting themselves as reliable developers of topical concepts, in both cases we can observe how the artist is no longer necessarily seen as a producer of rarefied and *personal* products – but as a purveyor of branded goods.

The wind tunnel on the roof doesn't fit easily into any category. It's too roughly built to function as an actual research instrument, yet it is not presented (not *selected*) as a symbolic object. It is neither branded, nor neatly tied to a specific theoretical discourse. Its purpose is not clearly stated, and this, I believe, is an important

part of its raison d'être. The team responsible for its existence – a group of artists and academics active under the banner of 'trans-disciplinarity' – have indeed built it all by themselves.

Some time ago, I attended an artist's talk with Vija Celmins, whose obsessively handmade and Zen-like paintings and prints I have admired for a long time. I was curious to see the person behind the work and did not really know what to expect, yet Celmins surprised me with her attitude. She simply refused to be drawn into any sort of theoretical or philosophical speculation. At one point her interlocutor tried to coach her into saying something about her work's relationship to the sublime. Celmins: "I don't think about the sublime, no." A little later, she was describing what makes her know a work is good: "You look at a work of art and you feel a little 'boing', a little surprise – or not".

When I am out on the green field striving to optimize my model's performance and my ability to read the atmosphere and fly it, "little surprises" are the last things I want to experience. My aim is the opposite: all my effort is directed towards the (impossible) goal of surprise-less perfection. After performing a successful catapult launch into the right air I experience no little "boing" but instead a sense of relief: "I did my best and what happened, happened just as it should." This is the essence of sport: forever trying to eliminate mistakes and prepare for perfection.

In the half century that followed after the Wright Brothers' wind tunnel, ever larger and more powerful constructions were built in order to support the astonishing development of flight and aerodynamics. Yet today, the development of dedicated mathematical theory[10] combined with the availability of vast computational power makes it possible to perform wind experiments virtually. Building wind tunnels may no longer be necessary for science. A practice of trial and error is giving way to speculative computation. More and more research today is migrating towards a fictional realm,[11] in an incessant march towards more detailed and refined prognostication and interpolation based on boundless collections of data.

In art, concurrently, anything may be assigned a value, using the mechanisms of naming and selecting. Where do we see something akin to basic research in art today, when on the one hand the principles of the market (entertainment value, saleability, scalability, branding) and on the other the theorization (virtualization) of

10 I.e. Computational Fluid Dynamics.

11 Research on perfecting LDA airfoils for use in F1A models has also been conducted using virtual wind tunnel software, such as the application "XFLR5", written by André Deperrois, by which the viscous drag associated with the wing can be estimated in 3D (see Bognolo 2011).

content-related matters (symbolic usefulness, verbalization, explanation, academization) influence all aspects of art production? When the artist's *making* in so many instances is subsumed to somebody's discursive construction and the Duchampian practices of *naming,* which allow retroactive adjustments to the work's construction,[12] not to mention the market's all-encompassing appetite for branding? One response has been trying to merge art practice with forms of science and its claim to usefulness. Perhaps we might even see a shared interest in surprises come to the fore again?

Could the wind tunnel on the roof be seen as an act of resistance, a bold alternative to the de-emphasizing of *making* (it yourself) in art and science? During a presentation, I hear another visitor's[13] reaction to it quoted: "Is it art? What else could it be!"

Your reaction to this statement will depend on where you come from and on where you want to go.

> Ever tried. Ever failed. No matter.
> Try again. Fail again. Fail better.
> (Beckett 1983: 7)

When we perform sport, we do it with an underlying wish to confirm ourselves to ourselves. We wish to perform to the best of our abilities within a chosen field, but not generally to extend it. Focusing on graspable challenges, for the promise of short-lived joy and honour, is a way to manage, guide and experience one's existence, but not to question it. Making art, on the other hand, will (or do I need to say 'should'?) involve looking to see oneself from the outside and ultimately to escape the self's restrictions. It has been a truism that you make art to know yourself, but in reality you do it to know yourself *as an other*. "Je est un autre",[14] Arthur Rimbaud wrote in his famous 'Seer's letter' to Paul Demeny,[15] and then continued a little later: "J'assiste à l'éclosion de ma pensée."[16]

I'm staring into the sky, focusing on the model which has continued to soar. During a flight in thermal the model might reach considerable height and drift far away downwind. Depending on the circumstances and the planned length of the flight, I will use my naked eyes or watch it through binoculars mounted on a tripod. No flight is similar to another, there is always an element of unpredictability, a risk that the model will fly out of the thermal, or conversely: that it will find thermal on its own when released in

12 A practice indeed pioneered by Marcel Duchamp.
13 Pavel Büchler
14 "I is someone else."
15 Dated 15 May 1871. Available online at: http://www.deslettres.fr/lettre-darthur-rimbaud-a-paul-demeny-dite-lettre-du-voyant-je-est-un-autre/ (accessed 9 February 2016).
16 "I witness the unfolding of my thought."

References

Beckett, Samuel, 1983, *Worstward Ho*. London: John Calder.

Bognolo, Claudio, 2011, 'F1A 3D Analysis with XFLR5' in *NFFS Symposium Report*, published by Free Flight News.

Goldstone, Lawrence, 2015, *Birdmen*. New York, NY: Ballantine Books.

Groys, Boris, 2005, 'Multiple Authorship' in Vanderlinden, Barbara, and Elena Filipovic (eds.), *The Manifesta Decade: Debates on Contemporary Art, Exhibitions and Biennials in Post-Wall Europe*. Cambridge: MIT Press: 93–100.

Oxford Dictionary of English. Oxford: Oxford University Press. Online at: https://en.oxforddictionaries.com/ (accessed 10 March 2017).

Sweeney, James Johnson, 1946, 'Eleven Europeans in America' in *The Museum of Modern Art Bulletin* 13: 19–21.

less than perfect air. One characteristic of a perfectly adjusted model is that it will hardly circle at all in neutral air, yet will begin tight circles of its own accord once it has detected a thermal. It's a beautiful sight to behold.

Studying my model's unguided flight I deduce which way it is going, while also keeping an eye on the distance flown in relation to the size of the field. I don't want to land in a tree, or worse. At a pre-set cut-off time, the timer springs into action and issues its final commands.[17] The minute arm holding down the back end of the stabilizer – offset on a tiny wheel which has been executing its precise movements during start and launch – is flipped back by servo 1, so that the stabilizer, attached at its front to the fuselage's aluminium holder with a rubber band, snaps up to a negative angle of 30 (real) degrees, held there by a thin string. 00.20 seconds later, servo 2 moves the rudder to value 80.0 and at 00.30 seconds, servo 3 puts the right half of the wing at incidence value 70.0. The plane stops flying completely and enters a slow spin to the ground. Light as it is in relation to its size, it will not be damaged on landing, as long as it doesn't hit a stone or another hard object.

My binoculars feature a built-in compass. If the distance flown is long, I will note the bearing of where the model has come down and enter it into my handheld GPS. Additionally, I may use my small walkie-talkie as a scanner; it's set to the frequency of the plane's tiny radio beacon. I start walking briskly to retrieve the model. The distance from start to landing depends on wind and time flown. In still weather, there might be no distance at all; windy, it might be very far.[18]

It's a beautiful day with hardly any clouds. I have watched the model come down safely; it was a successful flight and I can relax for a moment and forget the competition I'm taking part in. I'm walking under the sky, a tiny speck on the surface of the planet. I'm alone. My thoughts are in the here and now: my eyes looking to spot the plane as a disruption in the repetitive vegetation. Right now that's all there is, and it is good. There is no need to *select* anything. Just find the model. Here: on Earth, beneath the heavens.

Look, there is something, over by the small bushes! A fleeting glint of something white and red has caught my eye. I walk nearer...where was it now?...yes, there it is. I have found my model.

Now return. Start again. Not fail. Win. Win again. Please.

17 It is also possible to send a non-reversible abort command by radio signal, the only remote control possibility allowed by the rules.

18 When average wind speed reaches more than 5–6 m/s, flying F1A should best be avoided.

Landscaped gardens

When I was a child, one of the highlights each summer was when we drove from where my grandparents lived out into the country to visit a castle with a park. We could view the castle itself only through a fence. Since my mother was allowed to sleep there once at a girlfriend's in her youth, we suspected that kings no longer lived there. And something else was wrong with this castle, too. With its oriels it looked too massive and playful given its fortress-like walls. But it was still a castle, and owing to its spatial distance and inaccessibility it was charged with mysteries. Each time, we lingered at the fence for a long time. That's how we got in the mood. Then we walked past the orangery, which sold coffee and Neapolitan ice cream, over to the beckoning park. Had the latter only consisted of avenues, situated on a hillside with grassy areas and a pavilion, we would have quickly explored it. It would have become the destination of boring walks. But this park had no fence and no boundaries. It spread out; its paths became more branching, the arrangement of the trees more unformed. Undergrowth mixed in, wild flowers and ferns. Increasingly the park seemed to become wilderness, but then you unexpectedly came upon an isolated field. We never saw farmers or any trace of agricultural work there. Behind the field there was a wooded incline leading up to a hilltop. In the forest there were no longer any feeding troughs, nor any mysterious wells or garden houses likes the ones we had seen sporadically between the trees previously. The woods behind the field looked like a separate realm beyond the park, spreading out into an unforeseeable breadth. But it only proved to be a passage, with the destination up on the hill: the ruins of a mediaeval knight's castle. In its condition with semi-intact rooms and incomplete walls it sparked our imagination. Every time, we stopped when we saw the ruins and experienced time in different dimensions for a few moments. Subsequently we climbed around on the castle and conquered it for hours on end. The afternoons invariably culminated in a picnic on the castle's roof that had been sealed with corrugated iron. From there, we could look out in two directions: on the one side, back at the enigmatically entangled park, and on the other at a vast expanse of forests and valleys extending all the way to the horizon. Initially we saw only pristine wilderness,

but then we always searched for a grey-brown dot amidst the green areas. We knew that there was another castle there, inhabited by a granddaughter of the famous Count Zeppelin. When we discovered this castle, we always asked someone to tell us the story about how in earlier times the count, when he flew by not far away upon returning from one of his trips, made a detour with his zeppelin solely to circle around these valleys and the castle. During these moments, the horizon of the park widened endlessly. We found ourselves in the middle of a landscape.

Garden design

A central element of the Japanese gardens from the Muromachi period (ca. 1336–1573) is their periphery, in conjunction with the view from outside the garden. Landscape Park Ramholz as I visited it as a child invites you to move around, in keeping with the English model. A Japanese garden, however, lends itself to deceleration or even makes you want to come to a standstill. In it, Nature does not open up when you hike into its vast expanse, but when you sink into contemplation of the innermost part of the garden.

The moss garden Saihō-ji (ca. 1339) in Kyoto was still designed such that visitors can wander through a maze of paths. Its small woodpost-lined trails lead over hills and bridges past teahouses and past a pond. In terms of its inner diversity, Saihō-ji is on a par with Park Ramholz. However, you do not get the urge to head for the next corner, but rather to stand still more and more often and linger. One's gaze pans over a composition of 120 varieties of moss. They interweave into curved surfaces, thus creating an image that appears as if you were looking from a bird's eye view at the timeless landscape lying below. But the image shatters as soon as it arises, because at the same time phenomena emerge that can only be recognized from up close: unevenness in the moss or slight movements caused by the wind or insects. Thus on account of the artful presentation two versions of Nature catch the eye at the same time: its timeless vastness and its momentary details. Your perception of far and near starts to oscillate.

While the moss garden engenders a twofold perspective of Nature through compression, the classical dry Zen gardens of the Muromachi era provoke a comparable perception through division. These gardens are no longer conceived so that visitors can roam around and discover things, but are geared to seated viewing. One looks from outside the garden at right-angled surfaces that are framed like a picture. While a picture refers to Nature in another medium, here Nature uses its own means to refer to itself. You view stone or pebble formations that rise up like hills from gravel surfaces, which in turn are raked to appear like frozen waves. As in Saihō-ji, this makes it seem as though you are overlooking a landscape

from a distant point removed from time. Yet as opposed to the moss composition, in this case the here and now does not occur within these structures. The stones and their wave pattern do not react when an insect flits over them, nor do they transform by themselves or vanish.

A key motif of Japanese aesthetics is to make Nature experienceable by applying a cut through its continuity. In dry Zen gardens, this motif is immediately accessible. By being outdoors, the gardens simultaneously show what they make absent, for the play of light and shadow falling on them emphasizes their motionlessness as well as the Nature that surrounds them. As a 'borrowed' landscape, its colours and articulations have a significant impact on people's perception of the rock gardens. Hence two versions of Nature are juxtaposed: a moving and an unmoving one; a momentary and a timeless one. Their relationship is seamless. The dry Zen garden cuts into the living landscape around it like an image of its death. With every articulation of the 'here and now' of the Nature that surrounds it, paradoxically a 'not-now-but-sometime' is articulated inside it. A momentary insight into its eternity is revealed.

Artwork

The castle park in Ramholz transitions from its landscaped part seamlessly out into wild Nature; a Japanese dry Zen garden, by contrast, turns away from it in the manner of a cut and, through stylization, permits another view of it to be experienced. A third variant is represented by the installation *untilled* by Pierre Huyghe, exhibited at dOCUMENTA (13) on the grounds of Karlsaue Palace. In this installation, fluid transitions are composed leading from a park to the interior of an artwork whose elements, however, include Nature itself. Huyghe's work lies in a place where it becomes apparent that the formed aspect of a park necessarily includes the unformed. If there had been no signposts to *untilled*, most of the visitors would probably have overlooked the work. You approach it via paths that branch off from the main avenue and become increasingly untidy. Later you come upon a terrain with rampant weeds, algae-covered ponds, a heap of earth, rubble, compost, a pile of walkway slabs and a tree trunk lying on its side. At first glance, nothing calls to mind art. But when you look again, you see on the ground a sculpture of a lying woman propped up on one arm. She could also be a stored-away item of the park, but a colony of bees wraps around her head. The unexpected interplay between cultivated nature and art immediately provokes both bewilderment and fascination. Now the eye has spied a focus within the unstructured-looking zone and lingers there. But with a third look back at what was previously only glimpsed in passing, your perception of the entire terrain changes. Suddenly everything seems to be

entwined in a network of aesthetic references. At the same time, further differentiations become visible: growing next to the weeds are toxic plants with a sign indicating that they are poisonous; on the ground near the tree trunk is an animal skull; two white breeds each of which has one leg coloured pink mix amongst the dogs of visitors; a person who seems to be a guardian of the artwork turns out to be the owner of the dogs. He ensures that they never leave the spot.

Although it is hard to find a common denominator between the different elements of *untilled*, this work does not seem arbitrary. The information that the tree trunk lying there is a chopped-down specimen of Joseph Beuys's *7000 Oaks* heightens this impression. In it, two perspectives of perception are condensed on the basis of which two interpretations run through the work like axes that connect all the elements. One axis extends from classifying something as pure nature to classifying something as pure artwork. If you string together all of the elements of the installation, you discover a scale with many different gradations. Starting from pristine Nature as dead (animal skull, tree trunk, compost) and growing (weeds), it extends to the planted (toxic plants), restructured (earth heap), bred (bees), all the way to the trained (dog), subsequently moving on to a person as trainer (of the dogs) and breeder and nature marketer (of the bees), continuing to objects (paving slabs), and finally arriving at the artwork. You might think that you have come to the end of a dynamic scale here, but at its end there are elements of the beginning and middle again. The sculpture is linked with the bred bees, and Beuys's tree represents both 'the pure artwork' and 'pure Nature'. A comparable scale can be extended from the past to the growing, the eternally renewing, the articulating in the here and now, the coming, right through to the timeless. Here, too, all of the elements find a category, and again the ends of the axes finally reconverge. Beuys's tree can be interpreted both as past living Nature and as art that can potentially outlast time, comparable with the sculpture of a lying woman which as a replica of a sculpture by Max Reinhold Weber (1897–1982) also marks past modernism. The work *untilled* by Pierre Huyghe deals in a complex way with how Nature and art defy clear classification and instead enter into an enmeshed relationship of indissoluble ambiguity.

Wind tunnel

When I first heard that someone wanted to build a wind tunnel at Zurich University of the Arts, I formed an inner picture: I envisioned someone standing in a wooden tunnel approximately two metres in diameter and eight metres long while a wind like that on the North Sea whistled around the person's ears. The inner picture

had something fascinating about it, for it suggested that an artefact wanted to compete with Nature. When I actually saw the wind tunnel for the first time later, it was disappointing in comparison with how I had imagined it. It was too small and not designed such that someone could enter it and be played upon by the wind from all sides. On the other hand, it was too big and winding to be viewed from a distance like a picture. Thus it did not lend itself to a physical experience or to contemplation. Instead, you approached it curiously, viewed and touched it from all sides, held your hand in an open aisle through which wind moved that was blown out and then sucked up again, and wondered how the whole thing worked.

A topos of modernity is to continually transcend what art is supposed to be. It is less about what an artwork negotiates inside than how it stretches the concept of art to the outside. One effect of this model is amazement. One was amazed that a black square can be perceived as a picture, a urinal as a sculpture, or a 4’33” pause as music. But up until past the middle of the 20th century, amazement was only a side effect of advanced works, for their central concern was to expand the sign system *art* both sensually and reflexively. In the meantime, however, it has become an option to define amazement as the main effect in contemporary arts. Works created using this concept aim at eliciting a feeling of being overwhelmed by the encounter with the inconceivable. Upon entering a completely dark room at the documenta, one wonders how it is possible for the crowd of people there to sing synchronously while simultaneously moving around without bumping into someone. And one is overcome with wonder upon encountering a simulation of the sun in a museum. But the astonishment is finite. It disappears as soon as you get used to the effect or understand how it is created. When after a ride on a ghost train a child sees a monster with its mask off smoking a cigarette, the child turns to other attractions. The secret of the dark room at the documenta is resolved as soon as you realize how bright it actually is and how simple the rules of the improvised singing are. And you cease to be amazed by the sun in the museum once you have become accustomed to its effect.

My inner vision of the wind tunnel anticipated the effect of amazement. But the picture fell apart not because I got used to the object but because the limits of my imagination were exposed. I had envisioned an artwork based on the notion of being overwhelmed. Instead I encountered an artefact that did not want to pull me away, but to throw me back: to the nature of my own imagination. The dis-illusionment did not lead to the end but to the beginning of my aesthetic and reflexive encounter with this construct.

Art, instrument

The wind tunnel, designed and built by Florian Dombois and his team, whose research at Zurich University of the Arts (ZHdK) focuses on transdisciplinarity, is currently on the roof of the university building. Today it is no longer considered fruitful to argue about whether something is art or not. Back in 1978, the philosopher and sign theoretician Nelson Goodman said that the question *What is art?* should be replaced by *When is art?* Hence whether something is classified as art or not depends on the spatial and temporal conditions in which it is presented. Artworks in the living room can be perceived as chairs, and chairs in a museum as artworks. But neither interpretation would have been conceivable 100 years ago. Furthermore, the spatio-temporal classification has been supplemented by a discursive one. A re-enactment triggers legal debates in a courtroom and aesthetical ones at a theatre.

Various wind tunnels used for aerodynamic research at scientific institutions are scattered around the world. The wind tunnel described here stands diametrically opposed to them. As a *transdisciplinary object* on the roof of an art school it does not position itself at the centre of *one* discipline, but at the intersection of all disciplines represented there. In this central position, it seems like an empty space between the arts, suitable for both their turning away from it and their turning towards it. In the first variant the arts safeguard, and in the second open up and reflect, the limits of their disciplinary self-definition in the face of their encounter with this unclassifiable object. In the case of an artistic approach, an encounter with the wind tunnel would be largely experimental and imaginative, as there would not be a model or a script for it. It would be an act of artistic articulation and imagination. You would have to at once move closer to it and further away, because it would be a matter of understanding what the wind tunnel is, on the one hand, and on the other of imagining what else it could be from the perspective of your own discipline.

At first glance, the wind tunnel could also be a sculpture, design product, or puppet theatre stage. These associations are reactions to its visual appearance. At the same time, however, it produces and moves something invisible, namely wind. This expands the spectrum of its possible classifications, for the wind tunnel could also set something in motion. In that case, its effect would be more important than its appearance. One possibility would be the production of acoustic vibrations and, related to this, of sound. The wind tunnel could also be an instrument.

It is quite possible that the first musical instruments in the world came about by chance, whereby a thing enabled a sound and a person got the idea to make music with it. There are comparable developments in contemporary music, where existing objects are

reinterpreted for musical purposes. The history of Central European instrument making, however, is one of purposeful design of sound generators in the context of musical visions. Instrument builders, interpreters and composers have for centuries inspired one another to differentiate and continually reinvestigate the relationship between instrument, musical structure and playing technique.

In this respect, the wind tunnel is not an instrument. No composer called for it in order to realize musical ideas; no interpreter asked for it to expand possibilities of sound or playing techniques. As opposed to objects reinterpreted as instruments, it doesn't even produce a sound. When the wind blows through it, it remains quiet. Its corpus could be used as a percussion instrument to elicit sounds with a stick, a super ball, or bow. However, that would mean bypassing the tunnel's beating heart, namely, its inner dynamic, in the face of the wind moving through it, the route it goes through, and above all its reversal from being blown to being sucked up. If one wanted to use the complex wind tunnel as a sounding body, then one would find that it already contains structured time, that is, composition. One would find oneself at a threshold: between a composer who images acoustic and temporal structures and an interpreter who articulates time that is already structured. Subsequently one might encounter the second meaning of the term *instrument*, namely, as a measuring device. It would show how the relationship between wanting to determine and being determined is shaped in the compositional process with this object. Furthermore, one might be reminded of a garden artist whose imaging of forms collapses into his reacting to forms because the elements he or she designs are also Nature. Taking this idea further, one might wonder what it would be like to incorporate Nature's dynamic ways in the act of composing, for example taking up animal movements as temporal structures. Finally, this reflexive-imaginary obstacle course might lead us back to 'classical' composing itself. One could encounter the question of how the relationship between wanting to determine and being determined also occurs in the compositional positing, for by structuring and shaping time while composing one encounters it again the next moment as a different time, namely, as a given form that calls for a reaction because the time of the composing self has moved on in the meantime.

As an entity between not-yet-instrument and already-work, the wind tunnel invites us to reflect it from a compositional inner view. Yet through this movement one simultaneously is transferred to a vantage point that commands an outer view of one's own discipline, because the wind tunnel pulls one into its own articulations. These generate different questions from those bound to a specific discipline. Presumably approaches from other disciplines would lead to similar movements. The outer view evoked by the

wind tunnel, in which the arts would find themselves, would be the place of transdisciplinary perception. Only by choosing an approach from a disciplinary perspective does such a transdisciplinary perception come about.

Garden or landscape?

The wind tunnel on the roof invites us to engage with it in an artistically creative way as well as reflexively. It also invites us to approach it contemplatively. It is a visible construction that moves something invisible. To this end, it grants tactile access in one place. Where the wind is expelled from an exit and flows on until it is sucked in again at an entrance, there is an open space. In this space, one can feel the stream of the wind. Yet it feels no different than any other airstream.

There are Japanese gardens from the Muromachi period that within their borders present the same trees found in their immediate surroundings. Due to the separation and the resulting compositional difference, however, they create two views of Nature. Owing to their stylized and clear arrangement compared to the wildly growing trees in their surroundings, the trees in the gardens seem abstract. The dry Zen gardens heighten this double perspective to the extent that they also incorporate surrounding Nature. As 'borrowed landscape', it shows what scarcely exists in gardens any more, namely Nature in its continually renewing appearance. In contrast, the gardens themselves emphasize their timeless dimension.

The relationship between the wind outside and inside the tunnel is comparable to that between a 'borrowed landscape' and a dry Zen garden. You stand in front of the tunnel and are played upon by the wind. At the same time, you know that a second version of the wind unfolds inside it. It could be made visible through smoke or flying objects. Windows have been installed expressly for this purpose. In this case, you would be both interpreter and agent. When approaching it contemplatively, however, you let the channelled wind remain invisible. You pursue it the way it is.

The reversal from outer to inner wind occurs in the form of a cut into its natural flow. In this respect, the wind tunnel is akin to dry Zen gardens, as the latter create a similar relationship between wild Nature and its artistic rendering. An onlooker can ponder this difference in the gardens. But in the wind tunnel, there is nothing to look at, because the wind is invisible. Thus the break between the outer and inner wind is supplemented by a second dimension. It constitutes a reversal from sensing to imagining.

In his book *Aktive Passivität: Über den Spielraum des Denkens, Handelns und anderer Künste (Active Passivity: On Margins of Thinking, Acting and Other Arts)*, Martin Seel discusses how from

References

Seel, Martin, 2014, *Aktive Passivität: Über den Spielraum des Denkens, Handelns und anderer Künste*. Frankfurt a. M.: Fischer.

antiquity to the 20th century the relationship between art and Nature developed, despite profound shifts, as one of model and replication (Seel 2014: 229). In contrast, today one sees oneself in a complex state of 'infinite entanglement' between Nature and culture. Martin Seel puts this transformation in the context of changing awareness in the face of ecological crises. In his view, art can be an aesthetic arena for reflecting on this radical change (Seel 2014: 223ff). The changed relationship to Nature can be described in a similarly conclusive way from the context of art itself. Since its detachment from an immanent logic of progress, art has challenged artists to re-examine with each work the question of its localisation as well as its interplay with the outside. For the entwined relationship between man and Nature in our present, Martin Seel chooses the image of the landscape. In his view, a landscape experience does not depend on whether the landscape is pristine, but whether one finds oneself "amid a wealth of phenomena" without a "distancing overview or panorama". According to Seel, in such a situation we realize that "in everything we do we move in spheres that we know always exceed the horizon of our perceptive faculties" (Seel 2014: 234ff, original in German). Within the framework of our childhood, this experience came about in the form of a reversal from the perception of a park to that of a landscape during our summer visits to the castle park in Ramholz.

Pierre Huyghe divided up the complex relationship between art and Nature in the interior of his work. By composing gradual transitions from what is built to what grows, and from the momentary to the timeless, and in the process providing polyvalent correlations of both art and Nature, he managed to elicit within an artwork a landscape experience of unfathomable entanglement. As opposed to the work of Pierre Huyghe, the wind tunnel shows nothing in its interior. Yet it also composes a double perception, through the difference between the wind as a natural phenomenon and as an aesthetic entity. In this respect, it could be considered an invisible garden. But since there is no such thing as an invisible garden, its imageability is infinite. The self finds itself in a landscape consisting of the interplay between contemplation, reflection and imagination. They become enmeshed and move into an open realm.

English translation by Burke Barrett

The wind tunnel emerged due to a well-defined concern. The artist and researcher Florian Dombois was offered a position at Zurich University of the Arts (ZHdK) in 2011 to develop transdisciplinary research. He accepted it under the condition that the research project could remain small and just serve as a model. And since as an artist he does not think in terms of themes, but of spaces and people, he invented the wind tunnel. With this instrument, models, scaling, transdisciplinarity and the interfaces between art and science are to be tested. Dombois convinced ZHdK's management of the necessity of the wind tunnel as an artistic laboratory, and in 2012 the Swiss National Science Foundation ensured him funding for the project 'Size Matters: On Scale and Size of Models'. With Reinhard Wendler as model theoretician and the composer, sound artist, and designer Kaspar König as wind tunnel builder, he put together a team and soon launched an international tender for two positions for doctoral students which in a selection process were given to Sarine Waltenspül and Haseeb Ahmed.

The wind tunnel is now the centre of a large-scale infrastructure at ZHdK and as an artistic laboratory has no predecessors. It is the invention of Florian Dombois, who wants to establish a new twist in the interplay between art school and research. Now faculty members, artists and scientists are joining forces to carry out experiments with the wind as medium and to research and reflect the potentialities of the wind tunnel in a transdisciplinary manner without having to address the relentless questions about the commercial viability of such a machine.

Today the wind tunnel stands on the roof of ZHdK in a place that was initially planned to accommodate an open sculptors' studio. But there are not many sculptors today, so the machine was able to occupy the spot, which makes a lot of sense. After all, isn't the wind one of the universe's most powerful sculptors? Tailored by Kaspar König to the needs of its users and custodians, the wind tunnel arose as a hybrid between a closed wind tunnel based on the prototype developed in Göttingen in 1908 and the open wind tunnel type invented by Gustave Eiffel in 1905. The Kaspar König model thus is also a facility relating to the history of science and does not have to meet the most advanced technical demands made by industry. No one expects the precision and sharpness required

in a specialized laboratory. Since 2012 the wind tunnel has served as a multifunctional form-finding instrument that provokes inventiveness and coincidences and inspires philosophical thought, which makes it a metaphor machine. The latter appeals above all to people with little knowledge of physics and aerodynamics.

Florian Dombois explains his view of beneficial transdisciplinarity and productive knowledge transfer by citing the example of the so-called man engine, another metaphor machine. In the nineteenth century, this mechanism was used to assist the journeys of miners in the Harz region up or down mine shafts that were hundreds of metres deep. It consists of two adjacent rods driven by a water wheel that move in opposite directions and to which foot platforms are attached. The miners had to perilously move sideways back and forth on these foot platforms countless times to arrive at the bottom or the top. Thus Dombois suggests that in transdisciplinary exchange one should switch back and forth from one discipline to another to arrive at extended and more profound levels of experience, knowledge and insight.

The wind or the air, respectively – just like the three other basic elements fire, water and earth – have had fundamental importance in rituals and myths as well as in art and technology since time immemorial. Over the millennia, there have been streamers, banners, windmills, weather vanes, and all kinds of wind chimes, including ones that emit sounds or acoustic symbols.

The Futurists zealously explored the possibilities of measuring aerodynamics in the budding wind tunnels to find the ideal streamline and to optimize the forms. The aim was to speed up the pace of life and the development of modernism. In the following, however, examples from art will be cited in which artists directly and physically bring air, wind, or breath into play in their works.

Among the most important classical works of this kind are the mobiles of Alexander Calder (USA, 1898–1976), most of which move only slightly in gentle airstreams, and the kinetic outdoor sculptures of George Rickey (USA, 1907–2002), which in all kinds of wind and weather conditions yield to chance and the laws of equilibrium. In the works and writings of Mario Merz (Italy, 1925–2003), the poetic allegory of “vento preistorico dalle montagne gelate” (prehistoric wind from the icy mountains) emerged starting in the 1960s. With it, this important representative of Arte Povera refers to the journey of the wind through times and spaces; he calls himself a nomadic artist whose works embody both history and future potential with their everyday, inconspicuous, ‘povera’ materials – clay, wood, fruit, vegetables, canvas, glass, steel, colours, newspapers and neon.

Merz was fascinated by the paintings of cavemen. In his work *Irritato irritabile* from 1979, he refers to the prehistoric pictures of

hands that the painters created by spreading out the fingers of their hand (right-handers usually used their left hand) and pressing it against the rock like a stencil, and subsequently blowing the pigment over it through a hollow stalk. The result was an authentic document of individual existences, produced with their own breath, which have been preserved for more than 40,000 years up to the present day. These prehistoric paintings full of symbolic content are the first evidence of the use of air as a medium in art.

At all art schools and in many art studios today, air compressors are available for sandblasting or spraying paint. One of the masters of the spraying technique is Katharina Grosse (Germany, 1961). Using a spray gun, she drives her art onto walls, ceilings, floors and objects, letting the finely regulated jets of colour follow the course of her gaze. Hence the picture does not emerge an arm's length away, as when one paints with a brush, but metres away from the artist's body, from an expanded overview with which viewers, too, ultimately encounter the work. Thus Katharina Grosse's overwhelmingly expansive larger-than-life paintings look as if colour hurricanes had raged, leaving behind their traces. Spaces open up, not illusionistic pictorial spaces, but ecstatic colour spaces, vibrant stimulus zones and force fields offered up as emotionally charged spaces of thought.

With a potent wind machine like the ones used in a wind tunnel or on film sets, Jeff Wall (Canada, 1946) composed in 1993 one of his sensational works, *A Sudden Gust of Wind*, a giant Cibachrome transparency mounted in a lightbox. As a model, Wall used a coloured woodblock print from Hokusai's *Fuji* series, the one with the depiction of travellers who get caught in a wind storm on the Tokaido trade route near Ejiri station. But the Canadian stages his picture in the exceedingly unromantic, flat countryside outside of his home city Vancouver and transfers the scene to the present day as a *tableau vivant*. Four actors, a woman and three men, brace themselves against the squall in contorted positions. The wind blows a bundle of paper out of the woman's folder and whirls it along with autumn leaves from two bending, sparse trees and the hat of a passer-by over the churning waters of a pond. But this picture of a moment of play of fickle Nature is hardly a snapshot. Wall collaged his composition out of innumerable digital photographs. The many shootings cost him a lot of time and effort. The set had to be put up each time anew, all the sheets gathered, and the figures positioned properly again. But it was only during post-production at the computer that Wall gave the picture its final visual conciseness. Hence the work has to be read much differently than a snapshot. Every detail corresponds to the artist's creative purpose – Wall is extremely well versed in art history – as with

Hokusai or Paolo Veronese, resulting in a synthesis of film, photography and painting. Moreover, with the lightbox he makes use of the means of advertising presentation. But the main actor in the scene is the mechanically generated wind, which seems to cause the ideally conceived dramatic situation in this bleak landscape in which everything – the fields, the excavated pond, and the planted trees – is characterized by artificiality.

Another powerful art work created using air or wind as the medium is *Nothing* by Pipilotti Rist (Switzerland, 1962). The work was exhibited for the first time in the Arsenale at the Venice Biennale in 1999 and left a lasting impression on the audience. Rist devised a machine that was to emit large smoked-filled soap bubbles. Dimitri Westermann implemented the artistic vision. One after another, the soap bubbles follow each other in a slow rhythm, detach hesitantly and tremulously from the device, and buoyed by the wind sail majestically like gigantic baroque pearls over shorter or longer distances. In their shiny surfaces they reflect the iridescent colours of their surroundings in a distorted way, until the bubbles as an enchanting *memento mori* pop in a tiny and silent explosion; the small portion of captured smoke breaks out like a soul and soon everything dissolves without a trace, air into air. *Komm Schatz, wir stellen die Medien um & fangen nochmals von vorne an (Come honey, let's rearrange the media and start again anew)* is the title Pipilotti Rist gave to her large retrospective in Krems in 2014, an appeal that she herself rigorously complies with.

A final example refers to the category of acoustic wind chimes. In 2002, Pierre Huyghe (France, 1962) designed a wind organ edition for the art magazine *Parkett* that he called *All But One*. It has five aluminium tubular bells exactly tuned to the note sequence B – C – A – A (an octave lower) – E. This sequence quotes the iconic five-tone motif in Steven Spielberg's film *Close Encounters of the Third Kind* (1977) which helps five people and an extra-terrestrial make peaceful contact. At some point, according to the principle of chance, the pipes of Huyghe's wind organ might be struck by the wind in the right sequence to evoke interstellar communication again as a messenger. In his art, Huyghe creates adventurous events that to the viewers always become experiences which resound for a long time and trigger interaction – often with animals – as well as contingency and loss of control.

The few examples mentioned intend to show the important role ascribed to wind (air and breath) as an anthropologically-based medium and poetic instrument in art and in the constructed cultural realm of experience. Thus I see Dombois's idea of providing a wind tunnel at ZHdK for artists and scientists, as well as research

References

Schiller, Friedrich, 1910, 'Letters Upon the Aesthetic Education of Man' in Elliot, Charles W. (ed.), *Literary and Philosophical Essays – French, German and Italian, with introductions and notes*, vol. 32 (The Harvard Classics series). New York: P. F. Collier and Son: 221–320.

artists, who in their person amalgamate the two former figures, an absolutely ingeniously recognized necessity. The machine offers the opportunity to work in a simulating, intuitive, synthetic or intermedial way with the controlled airstream while enjoying a great degree of freedom. In practical terms, it can be used for the production of art works, and can also spark people's imagination to explore transdisciplinary reference fields.

Owing to its moderate size it would fit on the set of a chamber play, and its technical equipment is not too complex to be understood easily by a layperson to whom the facility is explained. It doesn't seem to pose any dangers. Two people can sit comfortably at a small table in the tunnel and drink tea while exposing their bodies to the airstream. In doing so, they can experiment with their sensory perception, test their behaviour and actions as well as their emotions and mood changes, and analyse the associations and memories triggered in them. Such experiments can also be witnessed by a small audience, for the wind tunnel has windows, and there is even a ladder leading to the roof of the machine, from where the goings-on can be followed from above. So even as a mere spectator who was not standing in the wind tunnel, I was inspired to ask questions such as: How can I strengthen my individuality so as not to blow with the wind when I stand amidst the crowd? How can I catch the wind figure of Kairos, the Greek god of the favourable moment, who has a shock of hair on his forehead but the back of whose head is bald, which means that I can only seize the opportunity from the front by grabbing his hair?

Anyone who engages with this inspiring wind tunnel facility invariably belongs to the type *homo ludens* and knows that culture-forming elements unfold and insights can be gained in games. Friedrich Schiller's remark in his philosophical letters on the aesthetic education of people still applies: "Man only plays when in the full meaning of the word he is a man, and he is only completely a man when he plays" (Schiller 1910: 266, Letter XV). In the culture of play, usually several participants are included. This leads away from the auratic work of the individual figure and towards products that only collective forces can generate.

English translation by Burke Barrett

Blowing in the Wind:
The Wind Tunnel as a 'Strange Attractor' of Artistic Research
Dieter Mersch

1.

The wind tunnel is a research machine. As an apparatus for manufacturing and controlling wind, it belongs in a larger ensemble with other technical devices used to examine the flow properties of technical things such as aircraft and cars, as well as bridge constructions, high-rises and the like. The actual machine is conceived as a circular device to enable accelerations; it has blowers, motors, energy supply systems and impermeable environments that protect the device from outer influences. Two functional principles can be distinguished: sucking in air and reinforcing its flow, and blowing and compressing air in order to aim it at appropriate speeds at objects placed exactly in the middle of the wind stream. In the process, the twofold character of tunnels is brought to bear: on the one hand the vectorization of currents, and on the other their compression and acceleration. The tunnel is used to measure different parameters such as air resistance and flow conditions, or the development of turbulences and their physical effects on the elasticity of selected materials and objects. However, the scale is decisive. On the one hand, aerodynamic properties can be tested based on real objects, in which, as with all scientific experimental forms, the restriction of the initial conditions and the elimination of disruptive marginal phenomena is pivotal. On the other hand, the vast majority of the studies are carried out based on models. In reduced form, the latter are subjected to 'real' wind forces, with the 'ruling' issue being the scalability itself – or to be more precise, the appropriate balancing of the relations between the miniaturized prototypes and the forces and speeds they are subjected to.

Thus we are confronted with a complex system of different dimensionings whose central problem is the asymmetry of the relations. For fluid substances such as wind or water cannot be miniaturised; they always 'operate', as it were, under real conditions, whereas some of the objects tested are subjected to considerable scale changes that generate their own effects. Reynolds numbers are used to balance this out, but they do not guarantee that the effects of models can consistently be transferred to their reference objects. What applies on a smaller scale is not necessarily valid for normal dimensions. Rather, scales have their own pitfalls

and unpredictable factors: storms in a water glass behave differently from real storms. Hence it is vital that the proportions and calibration are correct, but these in turn depend on countless measuring processes that make the required forecasts possible in the first place. The procedure is iterative: before an experimental setup is made, measurements have to be taken whose results inform the construction of the facility and the results it achieves, and equally serve to appropriately scale and control the facility and the relevance of the results it yields. The more precise the methods are, the more exact the research instrument becomes; and the better the latter functions, the more exact the results are, which in turn are used again to adjust the instrument and control its results. Hence the image of the wind tunnel is akin to that of a drawing where every new layer of drawings erases the previous imprecision in order to increasingly approach a mathematical diagram. Its ideal is a pure diagram not distorted by any materiality. Here everything is in the service, so to speak, of a comprehensive standardisation process: the closedness of the apparatus, the test gases that are already models themselves and circulate freely without being disturbed by outer influences, the circular course that ensures a homogenous airflow, the controllability of the pressure conditions right through to the continuously adjustable wind speed.

In scientific and technical experimental systems, a step-by-step rationalisation of all equipment and objects is of fundamental importance; what is more, the essential consequence of their statistic foundation consists in producing norms to begin with. Valid values can only be achieved through repetition under the exact same premises, whereas conversely the repeatability is tied to technical-mathematical structures, which have already implemented the machine and its mechanism, as it were, on every spot of the *dispositif*. Hence not only the device itself is machine-like, but also the underlying procedures, the tested prototypes, as well as the results, which consist of a series of data that have the form of mathematical functional elements. The data only seem to be valid when they can be reproduced and confirmed under the same prerequisites – in other words, when they have become signs stripped of every distinctive feature. As signs – Bruno Latour calls them "immutable mobiles"[1] – they can be transported, set off against something, cited or operationally manipulated and processed into graphics without consideration for their respective phenomenality. Indeed, the in part considerable technical elaborations and restrictions have no other purpose: they follow the logic of an identity production that applies the machine to all its parts and subjects the arsenal of media to a technical a priori everywhere. Consequently, science is mathematics, is semiotics, is technology. It transforms the world

1 See e.g. Latour (1987).

into an arsenal of signs that can be processed by formulae. The wind tunnel is its paradigm and reflection object. It investigates the wind as a statistic entity as well its effect on the behaviour of technical materials and things; and it does so on the basis of technologies which, in turn, do nothing other than optimize those technologies on which they are based.

2.

Art contradicts this hypothesis. It traverses the technical nature of the instruments, their scientific 'mechanics', their statistic preparation and its 'tunnel vision', in order to visualize things that are concealed in the technical arrangement. Thus it caters to reflection on the mediality of the technical nature of the instruments. Technology, however, whose meaning once included art in the word *technē*, exhausts itself in its functionality, its purpose. It is blind to its own functional status. Consequently, while the conditions for the effectivation of the mathematical basis of functions are inherent to the functions, a different view is required, a shift of axis, so to speak, to see these functions themselves. Wittgenstein expressed the change of perspective in his *Tractatus logico-philosophicus* in an inimitably simple and formalistic way:

> A function cannot be its own argument, because the functional sign already contains the prototype of its own argument and it cannot contain itself. If, for example, we suppose that the function F(fx) could be its own argument, then there would be a proposition "F(F(fx))", and in this the outer function F and the inner function F must have different meanings; for the inner has the form φ(fx), the outer the form ψ(φ(fx)). Common to both functions is only the letter "F", which by itself signifies nothing. (Wittgenstein 1955: 42)

But in order to write such a sentence, one must have already gone beyond the syntax of the functions, viewed them from the outside, so to speak, and noted their difference to their medium. To this end, reflection is required that is both philosophical and aesthetic. In addition to philosophical concepts, art is also a tool for it. Yet reflexivity suspends the technical functions at times, undermines them, reads them transversely to their actual destination, in order in the very first place to be able to make them visible as functions. Art is one of the means that can be used for such a visualization. Artistic research occupies a special place in this endeavour: it uncovers the invisible, in this case the entire framework of the scientific milieu and its apparatus, in other words that which one could call their mediality. So what does the wind tunnel mean then as an art object? In a technical-scientific context, it is

an attractor for generating measurement data, but as an art object it would be the exact opposite. This has a number of consequences, which at first may appear strange, but their plausibility is revealed on a second, delayed level. It includes the transition from an "epistemic thing" (Rheinberger 2006, original in German) to the 'aesthetic', yet the 'attraction' and special feature of the wind tunnel as art object lies in its being epistemic *and* aesthetic at the same time – or to be more precise: it is epistemic *as* an aesthetic object and aesthetic *as* an epistemic one.

For in the first place the 'strange' artefact standing on the roof of the ZHdK, which looks like an alien body that has landed there, is an installation, a piece of 'object art'[2], so to speak. The difference between it and the technical thing, the machine used to determine causality based on wind power, already reverses this perspective, converts the references, and reveals in light of the aesthetics the foreignness or alterity of the apparatus, its boldness and strange form, the monstrousness of the construction, its individual structures and thus its interior, how it functions, and so on. Then the 'aesthetic thing' can reflect the "epistemic", which certainly includes the provisional arrangements, the parts of the facility built of Plexiglas that provide a glimpse of what is otherwise hidden, the improvised aspect of the entire architecture, its obscene and exposed aspects, as it were. One of the early media-theoretical phantasms consisted in the literal 'disrobing' of the apparatus, the removal of the surfaces, panels, and non-functional décor, in order to expose the rods, pipes and cogwheels, in short the entire "frame-work" ("Ge-Stell", Heidegger). Fitting in the case of the wind tunnel is the disappointment when the clearness of the transparency comes about, for there is 'nothing' to see. The principle of the tunnel forms an empty space that envelops that which plays the main role but is not visible: the wind and its different speeds. Thus the choice of the tunnel as a paradigmatic artistic object has a deeper meaning: the technology used to test technology scientifically turns out to be null and void, an unknown that only heightens the mystery of the machine. As though by magic, in it effects and figures emerge from whose traces indirect inferences can be made which evade direct cognition. Hence the knowledge *dispositif* in no way functions out in the open, in the exoteric, in that which can be clearly seen by everyone, although there seem to be hard-and-fast rules, but – literally – speculatively: making the invisible visible.

However, in order to convert the machine, the technological object, into an art object and thus into a metaphor of a reflection of the machine itself, an 'imitation' (*mimēsis*) is required, its literal reproduction with all the redrawings of plans, the analysis of its

2 For a history of object art, see Rebentisch (2003).

architecture, the appropriate and inappropriate materials, the endless manual work with its errors, impasses and successes, until finally something emerges *that looks like* a wind tunnel yet actually *is not* one, but rather a singular thing, an observer or agent from another world that looks at the original apparatus from an oblique angle and through 'different' eyes. The special evidence of art, Adorno elaborated at the end of his Kranichstein lectures, consists in doing lots of "things in ignorance of what they are" (2003: 540, original in German). The art tunnel is an emblematic example of this. The ignorance lies in its 'speculativeness', which unfolds its special form of knowledge in the very first place in the prospective. It is what it will be. The epistemology of the aesthetic consists in its scope of possibility, and a criterion of art would be how wide this space can open up. Every detail is relevant for this. It already begins with the decision to set up the construction on the ZHdK's roof. In no way does it prove to be arbitrary: although the tunnel originally was built in the 'Kunsthof', or 'art yard', at the university's former location, it is a question of subjecting the tunnel itself – as a wind machine – to the wind, i.e. to integrate disruptive additional conditions rather than blocking them out, in order to make 'falsification' – or at least 'falsifiability' – part of its agenda. One consequence of this is that it is not about generating objective results and verifying them, about the 'definiteness of their truth', but on the contrary about unverifiable results in their irreplaceable singularity. One must also take into account the countless more or less successful or failed attempts to 'imitate' the functionality of the device, its adjustment and calibration to make it 'apt' for its tasks as precisely as possible, tasks which it ultimately does not fulfil, for the wind tunnel is not prepared to accept models for thorough testing, but rather the models act as representation aimed at investigating the tunnel itself. At the same time, one should not forget that the art tunnel is not an object of apperception that has a place in an exhibition, but an institute for carrying out real projects which – and this is literally what is 'clever' about it – risk their experimental character and precisely for this reason are able to make it visible. Thus a performative-reflexive note is attached to the aesthetic object, namely a moment of intervention of the 'art world' *in* the 'technical world' of the sciences, which is primarily characterized by the fact that the 'dis-placed' artefact brings about equally 'dis-placed' and self-willed attempts, with no recognizable 'purpose' whatsoever, except for the manifestation of their experimental character and its consequences. The privilege of art, its alibi, is its 'uselessness', whose oblique, secondary usefulness that comes into play 'sideways' consists in a smuggled-in reflexivity, as it were. Thus experiments can be generated without any apparent aim, like idle runs, which, however, reveal the specific poetics of the experimental: piano strings that are sounded by the

wind, thus transforming the machine, its technical nature and power, into an Aeolian harp, a musical instrument, right through to coal dust, which sweeps in its own, fleeting way around the corners of the tunnel, thereby leaving behind figures of special iconic appeal. For the art object 'wind tunnel', then, the way it functions and how to optimize it are not decisive, but rather the experience that can be had at the object itself and that relates to its manufacture as well as its contrary use – its 're-flection' on what it is: not a machine, but an attractor of aesthetic research, and as such, a "strange one." [3]

3.

This is reflected by the experimental character itself that is brought to bear in this way. It differs radically from scientific experimentalism. For when we examine the differences between artistic and scientific practices, it becomes apparent that the latter are result-oriented, while the former merge with the specific experiences arising during work. Consequently, the experimental in the scientific proceeds teleologically, but in the arts erratically and chaotically. One might think that something that wanders about intrinsically tends towards order, but at best it embarks on a search whose goal lies in the groping itself and in the process discovers something that has never been conceived of before. At the same time, it leads back to what is contained in the *experimentum* in the strict sense of the word: the *ex-periri* as an expertise that has matured through trial and error, and certainly also has connotations of being exposed and endangered.[4] It is this concept of *experientia*, an effort or undertaking that goes hand in hand with testing and even precedes and constitutes scientific experiment: an *exposing oneself* to the unknown, which only gains its 'knowledge' by keeping itself open to 'en-counter' and that which cannot be anticipated. *Experimentum*, then, means first and foremost that which, through such processes, has to be 'worked out' (*ex*) and 'expressed through the process' (*per*), which requires a passage that in no way prejudges what *reveals itself in the process*. Therefore the experimental is primarily about the exposition of events in their singularity, about that which comes towards (*Entgegenkommendes)* and its 'experience' (*Widerfahrung*), and only in a second step about something identifiable and manageable that increases knowledge. As opposed to the sciences, it is thus about difference and not repetition, about the other in the same, or attention to nuance, the evasive and non-conceptual. Also, while Heidegger

3 *Strange attractors* is a term used to refer to regularities in otherwise chaotic states and conditions: points of convergence towards which otherwise incalculable dynamic systems develop. The expression is used here as a metaphor for aesthetic research, which operates within the realm of the uncontrollable.

4 Cf. Vaniček (2005: 487f).

emphasized that "to set up an experiment means to conceive a condition" (1972: 74f, original in German) and thus to establish a law, a provision, aesthetic studies decontextualize and decontrol the contexts and their framework in order to open them up to the unexpected, the surprising. And instead of aiming at a growth of knowledge, an advancement of knowledge, they question their prerequisites to loosen them up a little and make them more uncertain, in order in this way to increase the room for sensitivity and conspicuity. Thus the actual aesthetic *experiens* always refers at the same time to the experiencing-oneself-in-the-experimental: artistic research progresses *esoterically,* when the word is understood in its original meaning of 'entry', a *return to that which is one's own*, and not in reference to 'exits' of the scientific public, its exoteric publicity and popularity.

So if we compare art and sciences based on an exemplary machine, as offered by the wind tunnel, we are confronted with two different research concepts and strategies. They can be subdivided into *ereuna* and *zētēsis* in reference to the classical Greek tradition of wisdom. While *ereuna*, and correspondingly *ereunaō* as practice, means inquiry through detection, *zētēsis* and *zētein* or *zēteomai* stand for a reflective search, for self-research. If we then have a look at their etymological roots, *ereuna* goes back to the extraction of knowledge from the still raw and unformed, whereas *zētēsis* goes back to the word stem *zē*, which roughly means 'achieve' or 'strive for'. *Zētein* and *zēteomai* are terms used for testing or investigating, as practiced by the judge, or the *zētētikos*, the philosopher who was sceptical in a special way and cultivated *zētēsis* as a practice. On the other hand, *ereuna exetasis* primarily means the inquisition, the method of 'interrogation', as embodied by the provocative claim that Francis Bacon made for modern research in general: namely, 'hound nature' until it betrays its secrets. In contrast, *zētēsis* denotes something completely different: not a teleological analysis that culminates in a goal, a result, but a 'pre-view' with no conclusion, as the tentative practice of permanent self-inquiry, which begins with the sensory, with seeing and hearing, and subsequently through perception and its medium reveals the whole circle of phenomena. It does not constitute research in the sense of *pursuit*, but in that of *passion*, in which it is not the *datum* that is the focus, but the ongoing doubt.[5]

If therefore the wind tunnel is referred to as a 'strange' attractor of aesthetic research, then in the sense of such an uninterrupted and simultaneously enquiring research process which at the same time does *not proceed investigatively,* in other words not in the form of a procedure, but always in the form of new and changing

5 It seems instructive that the *zētētikos* (zetetic, 'inquirer') emerges primarily in the context of the sceptical philosophy of Pyrrho; cf. e.g. Vorländer (1973: 152f).

References

Adorno, Theodor W., 2003, 'Vers une musique informelle' in *Quasi una fantasia, Musikalische Schriften 2*. Frankfurt a. M.: Suhrkamp: 493–540.

Heidegger, Martin, 1972, 'Die Zeit des Weltbildes' in *Holzwege*, 5th edition. Frankfurt a. M.: Klostermann: 69–104.

Latour, Bruno, 1987, *Science in Action: How to Follow Scientists and Engineers Through Society*. Milton Keynes: Open University Press.

Rebentisch, Juliane, 2003, *Ästhetik der Installation*. Frankfurt a. M.: Suhrkamp.

Rheinberger, Hans-Jörg, 2006, *Experimentalsysteme und epistemische Dinge: Eine Geschichte der Proteinsynthese im Reagenzglas*. Frankfurt a. M.: Suhrkamp.

Vaniček, Alois, 2005, *Griechisch-lateinisches etymologisches Wörterbuch (1877)*. Elibon Classics: s.l.

Vorländer, Karl, 1973, *Geschichte der Philosophie 1: Philosophie des Altertums*. Reinbek bei Hamburg: Rowohlt.

Wittgenstein, Ludwig, 1955, *Tractatus logico-philosophicus* (trans. C.K. Odgen). London: Routledge & Kegan.

'exercises' (*askēsis*), whose procedure consists in the simultaneity of a proceeding-against, a *reflectio,*– in short, an investigation that investigates itself and, in doing so, continually 'divides up' the perceptible (*partition*), as Jacques Rancière expressed it, in different ways in order to bring about new boundaries and new subdivisions or differences in its areas, so as to make things that have so far been invisible or inaudible visible or audible. This also means that research as a *zētētic process* is based, unlike scientific ereunistics, on a fundamental *openness to the unknown, the unexpected or the entangled, into which it constantly lets itself be drawn in and involved.* Consequently, its object is not the prototype, the model, but research and its medium, which relies less on the machine, on its technical equipment and potencies, than on the *wind*, which blows incessantly and simultaneously produces wasteful configurations out of itself, which counter technology (*technē*) with their own respective poetry (*poiēsis*).

English translation by Burke Barrett

Fahrkunst: On Working on This and the Other Side of Your Own World

Florian Dombois

And to each other they remain unknown, So long as they stand, the neighbouring trunks.

Friedrich Hölderlin, *Empedokles auf dem Aetna.*

A royal word is a thing… a thing… a thing that is nothing.

Georg Büchner, *Leonce und Lena.*

Many people, many voices. Many disciplines, little intersection. Art and science: you can put a horse and a narwhal in the same cage time and time again, but will the two really produce a unicorn? What do we actually expect when we talk about art and science, about artistic research or transdisciplinary research projects? What are the respective actors expected to do; what does working together actually mean?

After many years of crossing borders, after years of doing my own work in different disciplines, it is my conviction that the truly interesting aspects of a cross-disciplinary encounter do not lie in the search for a joint product or result. In fact it is not the others' knowledge that is the fruitful value of exchange, but the environment and the procedure from which it is gained. We have to visit each other at work, we have to work in the other's workshop then and again, where we often find the best things in the worlds of others in places where they don't look.

The approach presented here does not look for an interface between the scientific disciplines and various art forms, not a balance and not a mean value. It focuses, rather, on motivating experts to work within the framework of a condensed, parallel activity that brings a challenge for both sides. It is more about working in the same spaces and less about exchanging results. I want to instrumentalize the difference between disciplines; I want to work with their displacements. And not in the sense of a bundling of forces towards a common goal, but at best like with a twisted rope in which the fibres twine around one another, but do not run through from beginning to end. Or to put it differently: rather than thinking about transdisciplinarity starting from a problem or a thematic focus, I want to steer the attention toward the actors, and toward the spaces and time they dwell in together.

Let's imagine that a thinking style was a ladder. Let's imagine that the ladder can be moved within a disciplinary framework. Let's imagine a number of thinking styles and that the moveable ladders are installed close to each other. Furthermore, we would like to assume that a mind can linger in different thinking styles. How would an individual move meaningfully between these ladders?

Starting around 1800, mining pits became so deep that miners had to spend more and more time and energy going down and up them. At that time there were no lifts, among other things because there were no ropes of sufficient quality to raise and lower lift cars over great heights. The world's deepest mines at the time were in the Upper Harz mountains, with depths of 500 to 600 metres. And it was here that the mine mechanic Lichtenberg and subsequently the master miner Georg Dörell invented *Fahrkunst* (the man engine). The German names for the device itself and for all of its parts have the word *Kunst* in them, which meant 'technology' or 'art', in 19th-century German. This machine, which was operated from 1833 initially by water and later by steam power, converts the circular movement of the *Kunstrad* (water wheel) into a pushing and pulling movement, which is passed on via the *Kunstgestänge* (flatrod) to the *Kunstkreuz* (rotating cross-shaped lever). Hanging on this is the actual *Fahrkunst*, consisting of two ladders that move up and down in opposite directions between which miners had to switch in the right rhythm to ride down into or out of the shaft.[1] The respective ladders move only a few metres

1 An animation can be found at http://en.wikipedia.org/wiki/Man_engine (accessed 10 March 2017).

down and up, but by adeptly switching sides between the ladders the miner can cover great distances in a short time. So instead of being transported over the entire distance unmoving as in a lift, the miner in the man engine adds small movements on the two individual ladders to achieve one big one.

If we call it fidelity when one always stays within one's discipline, this procedure of changing sides is an infidelity. It requires self-discipline; it requires the courage to leave your own discipline, as well as to return to it. You have to hold onto the ladder but also to be able to jump back and forth.

You are tempted to do only one of the two: either to hold on forever, or to continually switch sides and thus to celebrate the undisciplined as an end in itself. *Fahrkunst* is temporary infidelity – for you return to the ladder you want to stand up for. To me, it is therefore a method to enable undisciplined thought, but not yet connected with the dissolution of disciplines. All the justified criticism notwithstanding, I like the idea of disciplines, for they are the places in which quality is struggled for.[2] Only the tightrope walk between the pitfalls of disciplinary narrowmindedness and transdisciplinary non-commitment makes *Fahrkunst* effective. The ladders and their mechanical movements can accelerate free thinkers unimaginably if they know how to move with and between them. This is not easy and makes many feel uncomfortable. In fact, this is often precisely the indication that one is on the right track: one no longer feels at ease in one's own skin, which is now that of another.

In the *Fahrkunst* of our project, visual and sound art came together; music and theatre encountered art history, culture and media studies. Accordingly, we did textual work and built models together. Articulations were made alternately in verbal and nonverbal formats, and we were also alternately required to prepare and contribute something written or an object. This changing of sides, which requires the effort of switching back and forth between language and nonverbality, is worthwhile and can be realised not only between different disciplines, but also within both scientific and artistic disciplines. The *Fahrkunst* presented here practiced the switching between thinking styles just as much as the switching between media.[3] This requires practice and concentration. So it is also good when one of the people in the group always provides the rhythm with which the group lets itself in for a different field and different forms of reflection each time. And several such movements are needed before one can articulate again with the sharpened precision of one's own discipline. And that was always the task of each team member: to make the respective monthly, joint working days fruitful for one's own work and then publish the conclusions drawn from it in one's own discipline.

However, there is a mixing of disciplines that notwithstanding all my above reservations appeals to me and to which we have tried to give a meaningful form with the *Wind Tunnel Bulletin* and the annual *Modellapéro*.[4] With this approach, I don't aim to create a publication as an artwork or as scientific research result, but rather a public place of energy transmission. I see these proto-disciplinary publications as heat exchangers with which something of the transdisciplinary heat of *Fahrkunst* is purposefully passed on to other professionals from various disciplines. Our bulletins are notifications, and our *Modellapéros* are casual get-togethers and not formal vernissages. I am not far along enough yet to describe these formats precisely and can only say that they intentionally remain 'unborn', always remain somewhat noncommittal to leave the greater part to the public. They are essentially generous, among other things because we are funded by the state, and thus committed to society in the best sense. And presumably they are research because others are needed to make a product out of them.

2 The concept of disciplines does not have a good reputation in the arts. And rightly so, for 20th-century art thrives to a not inconsiderable extent on the ever-new overcoming of disciplinary borders. However, this kind of borders do not exist any more today, in my opinion. Anything can be art. And that makes it possible to rethink the disciplines: in terms of their actors. So when I speak here of the discipline of art or music, I am referring to that group of actors who actively make an effort to produce art or music and who exchange ideas about what they each think are good or less good works.

3 "Mens et manus" is written in the coat-of-arms of MIT in Cambridge, to cite but one of countless examples that also proclaim a bringing together of linguistic and non-linguistic practices. And already in ancient Greek gymnasiums we find a combination of *res cogitans* and *res extensa*. So my concern to orchestrate verbal and non-verbal means is not new.

4 Cf. also the essay 'Forms of Publishing' at the end of this book.

In the following, the individual perspectives existing within my working group will be articulated. The articulations occur in language, because the book medium has been used. Thus something is said at least across one half. All of the texts begin the same way and then pick up steam in the direction of the author's own discipline. They are oriented to the respective peer group of their author and beyond it to a broader audience.

English translation by Burke Barrett

Taking the wind tunnel as a point of departure in its quality as an experimental system, a multitude of trails lead to the big and strange world of scaling effects. It is, one could say, an experimental system that makes these delicate spider web-like phenomena visible and palpable.

The following contribution is an attempt, based on the wind tunnel, to envision scaling effects in a bigger picture. This has long been difficult because, firstly, scaling effects appear in nearly all scientific and artistic disciplines; and secondly, because they are dealt with in various epistemic and pragmatic contexts in different, usually oversimplifying ways. Thus they are almost always understood as measuring and scaling *problems*. Consequently, their treatment is marked by a desire to avoid or eliminate them, and not by an attempt to probe them. But as strategies for solving the scaling effect problem belong to the multifactorial field of scaling effects themselves, they are part of the bigger picture and their genesis, role and effects have to be determined.

The streamlined form was essentially developed and authorized (Lichtenstein 1993) in the wind tunnel as an "experimental system" (Berz 2001: 394, 408, 420, original in German). On the one hand, this form served specific engineering-related interests concerning the acceleration of, for example, aircraft, trains, automobiles and projectiles. On the other, it is a symbol of the technological-scientific acceleration of the Euro-American cultural sphere, viewed by contemporaries simply as 'the world'. In the wind tunnel, 'the world' could be subjected to precisely those accelerated conditions that were diagnosed and thematized by the artistic avant-gardes from the 1850s until the late twentieth century. Metaphorically speaking, wind tunnels were used to test what kinds of things and which forms could withstand the new conditions. So the first wind tunnels were put in both scientific-technical and artistic-aesthetic contexts, although the latter were marginalised by engineers for various reasons. But this did not prevent the avant-garde artists from viewing the instrument from their own perspective.

The painting *Dinamismo d'un automobile* by Luigi Russolo refers in an exemplary way to the fact that the technical and aesthetic acceleration of the world did not stem from the wind tunnels, but was an essential condition for their development. The automobile

in the middle of the painting, which today is kept at the Musée Nationale d'Art Moderne Centre Georges Pompidou, stands for the technical aspect of this acceleration. It is overlain and deformed by six red-blue cuneiform shapes that are to be viewed as adaptations of supersonic paraboloids. They were first made visible by Ernst Mach and Peter Salcher's photographs of bullets. Taken around 1888, the pictures show a projectile and a Mach cone consisting of two shock waves. The latter arise when a projectile flies faster than 1200 kilometres an hour and, in the process, overtakes its own sound pressure waves, compresses them, and forms them into a cone.

Hence in *Dinamismo d´un automobile* Luigi Russolo imagined a supersonic automobile whose sound paraboloids penetrate the city discernible in the background. In this painting the automobile is not only shown to be a producer of sounds, but also of invisible cone forms, and thus at the same time as a musical instrument along the lines of the *Bruitismo* developed largely by Russolo, and as a sculptor of dynamic works in the sense of a Futurist view of art. In the Mach-Russolo reception relationship exemplified here, Futurists do not appear as 'salvagers' of the technological-scientific development, but as researchers who view a few parts of that broad field that was concealed from view by the pragmatic reductions of the scientists and engineers working in the wind tunnels.

Even today, there are aspects and meanings of wind tunnels that go far beyond the merely technical field of view. Thus a long series of different wind tunnels can be seen in TV commercials, not only for automobile companies, but also for postal delivery companies,[1] telecommunications providers, television manufacturers,[2] awnings producers[3] and trading companies.[4] In Paris Bois-Colombe, the Hispano-Suiza wind tunnel was converted into the École La Cigogne. Its spectacular suction opening is illuminated with red light in the evening. The large wind tunnel in Berlin Adlershof with its astonishing acoustics served as the setting for the films *Æon Flux* and *Wer ist Hanna?* and is used on many occasions for cultural events.[5] The big wind tunnel S1Ch in Chalais-Meudon, France, a Monument Historique de France, is used today as an exhibition space, and is to be converted into an event and conference room.[6] The Southern California Cooperative Wind Tunnel in Pasadena was transformed into the Wind Tunnel Graduate Center for Critical Practice and now offers space for many artistic and scientific events. The Old Farnborough Wind Tunnel of the Royal

1 https://www.youtube.com/watch?v=TATdv6_Zss0 (accessed 14 February 2017).
2 https://www.youtube.com/watch?v=kfPFjwASSIM (accessed 14 February 2017).
3 From 0:46: https://www.youtube.com/watch?v=Lvu2qorGF4s (accessed 14 February 2017).
4 https://www.youtube.com/watch?v=9yvVzpHl7Qo (accessed 14 February 2017).
5 Jazz im Windkanal, siehe: https://www.youtube.com/watch?v=XcDG5uDAIG8 (accessed 14 February 2017).
6 http://issuu.com/tiphainehsa/docs/carnet_n__3_-__projet (accessed 14 February 2017), p. 55.

References

Berz, Peter and Christoph Hoffmann (eds.), 2001, *Über Schall: Ernst Machs und Peter Salchers Geschoßfotografien*. Göttingen: Wallstein.

Lichtenstein, Claude and Franz Engler (eds.), 1993, *Stromlinienform, Streamline, Aérodynamisme, Aerodinamismo* (exh. cat.). Zurich: Müller.

Aircraft Establishment in the English town of Farnborough is used for artistic events and concerts.[7] The ACE Climatic Wind Tunnel of the University of Ontario is regularly used for concerts and student parties. The artistic wind tunnel of the transdisciplinarity research focus at the Zurich University of the Arts is used as an artwork, art space, instrument, performance venue, and much more, and at the same time discussed in terms of the multiplicity of its identities.

In such and numerous other forms, the wind tunnel is divorced from its pragmatic engineering-related simplification, as it were, becoming visible again in fragments of its cultural-aesthetic dimension. Hence it can be viewed as an example for many other places, objects and circumstances that are also purposefully reduced when viewed from various scientific, technical and economic perspectives to just a few aspects and thus tailored to the respective particular needs and aims. The dominance of these perspectives and the associated limitations of the aesthetic, visual and emotional competencies is unnecessary beyond the specific pragmatic contexts. Projects such as the artistic wind tunnel make it clear which paths lead out of the situation and what can be gained there.

English translation by Burke Barrett

7 For example, http://thewindtunnelproject.com (accessed 14 February 2017).

Taking the wind tunnel as a point of departure, I experience the essence of my artistic-scientific investigation in the tunnel as invisible energy consisting of wind molecules that compress and stretch. The wind is at the centre and is the trigger for our projects and their positive and negative lift. In and with the projects we are subject to poetry, romanticism and creative erosion, as well as fluid mechanics and institutional goings-on. In the wind tunnel, I turn on an airstream and then steer its flow into a movement through objects that are simultaneously tested in this artificial wind. So we work with wind both literally and figuratively. And even when an experiment is finished, I am driven by an invisible force that I can never really stop and that continues subconsciously. Then I take the flow and its effects with me for my next effort, keep them in my imagination to take the next step in the wind (tunnel).

This text attempts to describe the different relationships that I have observed in the wind flow and that continue to exert a subliminal effect on my research, and specifically also in my experiments.

Windy

As a member of the team and a designer of the wind tunnel, I am amazed at how many physical and metaphorical winds can blow around the wind tunnel at the same time. These winds (e.g. the wind on the roof and the wind of the institution) blow by themselves and can compress or stretch our work. And it is definitely windy up on the roof. Indeed, I am continually surprised about how and why exactly this or that is carried by the wind. And then I think about its poetic significance.

The wind seems like an energy or action that comes from off-screen, as it were. For example, issues relevant to safety repeatedly come up at our premises and become turbulent in and around the wind tunnel: examples are the escape route and emergency exit. Sometimes we have headwind, sometimes tailwind. Apparently, in the Toni Areal there was no other option but to expose the wind tunnel to the natural winds. Our research dwells there as though under a bridge. The sense of artistic research is critically questioned, and not only by the reflection group. These voices of the co-authors of this book, who are actually favourably disposed towards

our project, sometimes sounded like stormy criticism. What do we really want to show, in which area do we act? How can one work with the wind and why is a wind tunnel needed for artistic research? All of the questions around our process became part of the articulations, the working modes and our results. For what do we see when we look into the seeming nothingness of the wind?

Wind-me

In our wind tunnel, modularity plays a key role. It should be possible to convert and adapt it for every kind of test quickly and uncomplicatedly. This is an expectation I place on myself as a 'modern worker', just as I do on the wind tunnel. In today's world, modularity inspires me in many ways, and I think that more and more sub-products, components and sub-processes have to come together for us to survive. In this regard, I try to ensure that other users of the wind tunnel have an optimal workflow for their wind experiments; although in some areas a lot is missing for the optimal wind laboratory (e.g. an integrated smoke generating unit, or closing the cracks of the joints between the modules).

In our different tests, which also contribute greatly to my own performances (e.g. a concert at the ZKM in 2014 with a wind harp), or to my acoustic microphone experiments that will appear soon on CD, I appreciate the modularity and the 'hands-on' approach of our practice. The sound in the wind tunnel is particularly important to me. It is not easy to produce sounds in a natural way with air. For I do not simply seek an electronic set-up in which the wind merely changes a few parameters, but a truly acoustic set-up. Moreover, I seek a set-up that changes poetically with constant wind conditions and simultaneously shows its aerodynamic peculiarities.

I began with 3D printouts of model flutes and other aeolian sound generators. They became stimulating intermediate steps enabling me to make new 'discoveries': For through digital work with 3D prints, which can be scaled infinitely (and dimensionlessly), I discovered sound effects that I translated into real practice and could subsequently test in the wind tunnel. And from these analogue experiments I again drew digital conclusions that piqued my curiosity.

Wind-you

As an operator and artist in an institution, I try to get others enthusiastic about the initially invisible flow, about the tamed wind. The wind that touches you and me on the skin surface and has an effect without words.

References

Wind Tunnel Bulletin #5. 2016 (June). Zurich. Online at http://windtunnelbulletin.zhdk.ch/issue-05 (accessed 14 February 2017).

As soon one starts to articulate one's thoughts with words, one also articulates the wind's in-fluence, which I encounter with curiosity each time. We can perceive different colours – but what about the wind? When do you start feeling something… and when does *anemophobia* set in?

The wind between you and me will become an expression and lead to a relationship.

Some experiments only come about due to the emanation and the interaction with people, and I feel a great attraction to, but also a responsibility for, this reality. Owing to the encounters in the wind tunnel, I have built up a 'repertoire' of artistic procedures that sharpens my premonition of moments, permits me to mediate the invisible, and provides legitimation for this work 'without resistance'. A repertoire that engages with the wind and translates the invisible character of the wind itself into something palpable that does not stop at the skin. Far from it, alongside the sense of touch and sight I experience the ear as the most instructive organ of perception. And just as the hearing organ is directly addressed by a storm, I also experience lighter and artificial winds above all as an acoustic event. I, however, do research with great interest on acoustic measuring instruments, on new wind-hearing instruments. Entering into dialogue with tone and sound and inventing measuring instruments in the process appeals to me. But I can only convince you of this non-verbally. What do you hear?

Windproof

Our *Inflatable Wind Tunnel* (2015) has the form of an Eiffel-type wind tunnel and sucks the ambient air – or wind – into an enclosed measuring section. We use the resulting compressed air, which is usually blown back into the exterior, by redirecting it into windproof tubes that enclose and stabilize the measuring section. Hence the *Inflatable Wind Tunnel* combines tunnels or ducts that are open to the wind and windproof ones. And it enables us to perform location-independently. A large wind tunnel in a small transport sack.

In its form, the *Inflatable Wind Tunnel* calls to mind a calabash and the Eiffel-type wind tunnel. Due to the design, however, the outgoing wind of the test section flows back as feed wind, thus making this set-up actually a Göttingen-type wind tunnel. A suction test section that is in equilibrium with the inflated outer construction that gives it its form. Between sucking and blowing. The more wind tunnels there are, the fewer the free turbulences that can propagate.

English translation by Burke Barrett

Taking the wind tunnel as a point of departure, models are never far off. As building a wind tunnel and the respective models in full scale is an expensive and laborious undertaking, aerodynamics usually employs scale models. The phenomena that occur in this field in connection with the models used have been examined elsewhere. However, the use of models and model techniques is not restricted to research on wind tunnels. Models are also employed in other contexts to design, present, conceptualize, play, visualize, test and simulate, and hence there is a wide range of possible uses for and ways of engaging and interacting with models. Among these, I am interested primarily in those processes of visualization and simulation that utilize cinematographic models. In particular, I aim to examine the points at which aesthetically motivated and technically conditioned uses of models and model techniques intersect and overlap in the genre of film.

We encounter a wealth of cinematographic models in 20th- and 21st-century films. Sometimes the fact that they are models is disclosed, but often not. We can only speculate whether this game of 'hide and seek' is one of the reasons why models remain more or less invisible in academic scholarship. They are in fact well-nigh completely ignored in research, which is why a significant part of my work on models consists of basic research, such as compiling an overview of the (rudimentary) state of research and developing an appropriate terminology and theoretical and systematic foundations. This includes the analysis of the possible 'origins' of the various model techniques, such as (architectural) model photography or stage architecture, for example; or the historical overview of perspectives on cinematographic models or 'miniatures', covering the fields of technology, aesthetics, film history, the history of architecture, mathematics and physics. Here, my main focus lies upon the aesthetic and technology- and physics-related discourses that I consider of key significance if we are to gain a fundamental understanding of the models. These discourses must likewise be seen in close relation to wind tunnel research.

During my several years of work on cinematographic models, I have gone beyond basic research, producing a growing number of theses, some of which I will mention here. My main thesis is that models, which were first used primarily to create the illusion of or simulate objects, developed their own specific aesthetics

in 20th- and 21st-century films, an aesthetics I propose calling 'model aesthetics'. In this regard, I investigated various model techniques and their effects on filmic forms of expression and representation. For example, some films reveal that filmed objects are in fact models, as is the case in the 'Arctic World' explosion in Tim Burton's *Batman Returns* (US/UK 1992) or the flood scene in Wes Anderson's *Moonrise Kingdom* (US 2012). In both cases, hardly any attempt is made to hide the fact that the objects shown are in fact models. Techniques that otherwise are used to turn the simulative models into illusions are employed in such a way that the models and the associated techniques remain recognizably such: simulative methods become non-simulative ones. Furthermore, Anderson's films often give the impression of models even where none are used.

In a next step, I broaden my study of the (shifting) roles of cinematographic models to include an analysis of the way they are dealt with in narrative terms. One example of this is the scale model of a labyrinth over which Jack, the main protagonist of *The Shining* (US/UK 1980, director Stanley Kubrick), bends and in which he sees his wife Wendy and son Danny wandering around. Here, cinematographic model techniques are used to blur the border between models as motifs and models as illusions and simulations on both a technical and narrative level. This scene marks a key point in the storyline, suggestively introducing Jack's incipient madness.

Furthermore, I would like to focus particularly on the era during which it became possible to produce images using computer technology, investigating what happens to models from then onwards. Many contemporary films combine model techniques with CGI (computer-generated imagery). The decision in favour of or against one or the other is made based on what their use is meant to achieve, which is usually to create an illusion. Rarely, however – and these are the cases that I find particularly interesting – these techniques are not simply combined, but also *confronted* with one another, as is the case in the short film *Giant God Warrior Appears in Tokyo* (Japan 2012, directors Hideaki Anno and Shinji Higuchi), where both computer-generated images and model techniques remain recognizable as such.

In regard to all of these theses, it is important that the cinematographic models are not seen as isolated objects, but are instead considered in their relation to the respective model techniques and in their media-specific orientation, which is why I also focus additionally on the camera as a scaling instrument.

A solid foundation is needed when dealing with such a wealth of heterogeneous objects, which I have attempted to create by systematizing said objects. This system should be seen neither as absolute nor as complete. I categorize the cinematographic models as follows: (1.) models that simulate or create an illusion,

(2.) stylistic or non-simulative models, (3.) motivic models and (4.) transfilmic models and model techniques. I have already mentioned the first three types above; the fourth group includes cinematographic models and model techniques that have been removed from their original context – the specific film or the filmic medium – and have developed a kind of life of their own. In this category, I include models that appear as quotes in other films, either directly or slightly altered, and models that go beyond the medium and are translated into other contexts as merchandising products or exhibition objects. But not only the models themselves can be transfilmic; model techniques and model aesthetics can be, too. For example, the computer game *Lumino City* was filmed completely in a handmade model into which the characters were rendered retrospectively. Then we have views of cities, which are filmed with a special lens and from a particular angle or post-processed using tilt-shift effects so that they look like miniatures to our eyes owing to our visual habits, as in *Enter the Void* (France 2009, director Gaspar Noé).

My work on cinematographic models aims to provide a first approach to understanding these models not only as scaled, technical objects but also as mediated objects, which over the course of the past century have established a specific aesthetics that had a formative influence both upon film and beyond it.

English translation by Margaret Hiley

A Wind Egg:
We Aren't the Same and This Is Only Art
Haseeb Ahmed

Taking the wind tunnel as a point of departure I realize that we may not be the same and we certainly don't have the same experiences, but we can have experiences of the same *thing*.[1] Through the thing we experience the one who made it and so we can have experiences of one another, whether we know each other or not. We can also make something together to experience something together. This is a way of producing sameness.

Passing the Buck

Ideas pass from one person to another and through things that can become like people. This is the story of the Red-Haired Man. When I started as a researcher at the von Karman Institute for Fluid Dynamics (VKI) in October 2014, I took a Measurement Techniques course with the new engineering students. In the L7 wind tunnel, there was a flow-visualization experiment in which the turbulence forming off of a delta wing could be seen clearly when smoke was introduced into the tunnel and laterally cut with a laser sheet.

Guillaume Grossir taught that section of the course. Later Guillaume taught me how to fly sailplanes at a historic airfield in eastern Germany. Olivier Chazot, Director of the VKI Aerospace Department, and I would reconduct and discuss this flow-visualization experiment and perform it once more for Florian Dombois, my PhD advisor at the Zurich University of the Arts. Two years passed and I brought Marc Matter, a sound artist and my flatmate, with me to the VKI. He started to play with the delta wing as I watched. The stable vortexes shifted with the angle of attack of the delta wing model and then they appeared as eyes.

I made a video of the 'vortex face' and showed it to my studio mate and friend, the poet and critic Piero Bisello. He immediately thought of 'Red-Haired Man', a poem by the Russian futurist poet Daniil Kharms (2005: No. 10). This poem became a subtext. Piero was studying computer programming at the time and experimenting with algorithmic poetry, so he turned this into a script, at which I joked: "We can teach the wind to speak!" But what language

1 In conversation with Florian Dombois and paraphrased from Dieter Mersch: "We cannot assume shared territories, but we can have shared objects [between disciplines]." (Personal conversation, Uerikon, 27 October 2013)

does the wind speak? To find out, we decided to try four different types of language: randomly generated Latin syntax, the aforementioned poem, emoticons and hybrid live-typing. Piero wrote a new Python script and interface for this.

He added the output text to my video and I turned it into a scientific poster entitled 'Wind Personification', which I presented alongside other scientific posters by PhD engineering students at the conference *Drawing Conclusions*, held at the Swiss Federal Laboratories for Materials Science and Technology (EMPA) in December 2015. I was inspired by John Michell's book *Simulacra*, which is about anthropomorphism and seeing faces in nature. It was purchased on a road trip made to the John K. King Used and Rare Book Store in Detroit with my friend Chris Mansour, an enthusiast of artists' books.

Piero and I decided to perform the experiment, in order to teach the wind to speak for a live audience. Instead of using the L7 wind tunnel, which is fixed inside the VKI, I made a portable School-Lab Wind Tunnel (SLWT), which had been designed by the VKI's former director, Professor Carbonaro. At our shared studio Le Sceptre in Brussels, we previewed the performance for a select audience, including artists from the city and engineers from the VKI, like Olivier, as well as our friends Yakut Cansev and Ertan Umit, who would become collaborators on the project.

Ertan was so inspired by the performance that he offered to help us turn the *Red-Haired Man* (aka *Vortex Face*) into an animatronic robot. At the time of writing this text, in April of 2016, this robot is nearing completion. Using networked weather APIs the face will become the representative of weather patterns all over the world. All things that move must pass through a wind tunnel before they enter our industrially produced world; however, the site itself is rarely seen. It is therefore an other-space that is designed to reproduce any-other-space, yet it achieves this by creating a condition of laminar flow (straight-moving air) that exists in no other space. The SLWT relates to a particular regime of velocity, viscosity and density (Reynolds number). In it we have a particular slice of the world-wind, and we capture the face of the wind here to tell us about all the other winds in the world, which he does.

I say the wind is a *he*, because he has become a protagonist in a larger story I am concocting with others, the *Wind Egg*. Starting as an ancient Greek theory that animals and people can be fertilized by the wind in the same way as plants are, this theory was used for 2000 years, ending as an eighteenth-century, proto-scientific explanation for the Virgin Mary's immaculate conception. Yakut's experiment inadvertently tested the reality of a wind egg. Her master's degree project was a study on how much wind would penetrate a specially designed egg shape. The egg shape was placed in an aero-acoustic wind tunnel. Sound propagated in the

References

Kharms, Daniil, 2005, *The Blue Notebook* (trans. Matvei Yankelevich). Brooklyn, NY: Ugly Duckling Presse.

wind tunnel was carried downstream towards the egg shape. A microphone embedded within the egg shape measured the amount of sound, and hence wind, penetrating its surface. On 4 May 2016 this experiment will be part of a larger VKI project to resurrect the theory of the wind egg, with the aim of developing the possibility of human reproduction without men and with the wind. However, we will start with vultures, since the Greeks were certain that this species was exclusively female and procreated exclusively through wind-egg fertilization. Therefore, the wind is male.

The *Wind Egg* experiment will bring together filmmakers, aeronautical and aero-acoustic engineers, sound artists, vultures, a choreographer, a poet, actors and a falconer for an audience of engineers, the general public and the art community. For me the *Wind Egg* story started at one of our monthly 'Size Matters' meetings at the Zurich University of the Arts. Florian Dombois and Reinhard Wendler had an inside joke about a "classical wind egg." It came up again during one of our two visits to the Trudelturm (located in the Adlershof section of Berlin), a large vertical wind tunnel used to simulate the conditions of perpetual falling. It is shaped like a giant egg, and a constant wind once blew within it. The term *wind egg* emerged again and this time it stuck. Sarine Waltenspül found the technical drawings dating to 1933, and I used these to reverse-engineer and print a 3D model of the Adlershof wind egg. I also made it small in order to put my awe for the building into perspective. The Trudelturm is one of the best sculptures I've ever seen. Now, having researched the story and its Greek origins, I found its connection to another Greek use of eggs: egg-and-dart ornamental mouldings. I had previously used a plaster-replica of an egg-and-dart ornament in the first iteration of the project 'Has the World Already Been Made?', realized with Daniel Baird at the Jan van Eyck Academie in 2011. In 2014, for the exhibition 'A is for Albatross', I made an egg-and-dart ornament in which the eggs were replaced with the wind egg modelled on the Trudelturm's form, to enable me to bring inherited ancient and modern pasts together into a single form, in order to contemplate the future.

A Ride with Images and/or Objects? The Wind Tunnel as Site of Engagement

Julie Harboe

Taking the wind tunnel as a point of departure to reflect on materials, institutions and transdisciplinary collaborations, it becomes apparent how discourses, disposition or disciplines shape our approaches and understanding of these basic concepts. Working with the wind tunnel in the borderlands between art and scientific imagery positions us in an open laboratory studio. For the art historian, the tunnel is a tool and an aesthetic artefact. The project as such opens up a space of reflection without the dullness of convention, inviting drafts of different stories.

Welcome!

You are now in one of *Toni's talking transport lifts*. I am a relative of *Toni's small talking public lifts*. We move and talk all day and all through the night. That's right, not just when humans are around. I will take you from where you are to the eighth floor, the rooftop of Toni Areal. However, you can only come along if you hold, or are in the company of a human with, a Zurich University of the Arts employee or student card. Moreover, the card must be programmed to my use, in particular, of course. On the roof you will find the 'Size Matters' wind tunnel, aka 'The Temple of the Winds', aka 'The Air Dragon', aka 'The Metaphor Machine', aka 'The Instrument of Things and Sounds'. Heard all about it. I must stay here in the shaft, but will inform you of stops and movements on the way up and the doors will not open until I have told you.

Going Up – Globe

The image for this trip focuses on density and movement and connectivity of information. First, imagine academic production's endless interaction of impulses in one container. Somewhat like one of the fascinating snow globes you buy for decoration and play. You shake it and watch the movement of the flakes. Only this globe is full of swirling letters: footnotes and ideas, publications, words spoken in auditoria, in classrooms, on the street, in cafes, in solitude. Writings in computers and notebooks, photos and diagrams, which are all easy to transfer digitally as lots of zeros and ones. Now let the whirl in that globe comprise *only* the interaction with educational and research institutions, the academe,

no information from news or entertainment. This data is big enough. All we see when we focus on the snowstorm is trillions of connections, a density growing every second. We wonder whether the collective concentration, all the ideas, emotions and thoughts intensely interacting in what can best be understood as Brownian motion, will lead to smarter solutions for the planet, or just support its self-expanding mass. There is not unlimited space. The current challenge seems to be to find common objects to cluster and connect our activities in the twirl.

Eighth Floor

In a covered part of the roof terrace on top of the Toni-Areal we encounter the 'Size Matters' wind tunnel. It is made of wood, has a simple large motor with a fan that sucks air in, blows it out, and makes some sounds while doing so. Currently two parts of the tunnel walls are transparent. In the Eiffel test section you look into a tunnel that has a square cross section through which the wind is sucked. And in the middle of the wind tunnel a high window allows you both to look into a large space between two sets of guiding vanes and also to enter by pushing the window aside. The outside of the tunnel is the colour of glowing amber; the inside is painted white. The nozzle is beautifully swung metal and the long guiding vanes are also made from metal. Still, the wind tunnel is quite fragile. Like many wooden instruments it is not robust; though it functions it is just a model. Like a tree house, or a rowing boat, it was made for temporary and intermediary purposes. It works, but it would also be easy to destroy it. Handle it with care, but be free to test and use it as you please.

Welcome Back!

So on the roof you met the 'Size Matters' wind tunnel. They say it is a model, and that it is poetic but also silly, because it is neither tough nor high tech like we are. Certainly it is a bit of a nuisance to my colleagues. For instance, *Toni's fire alarms* and *Toni's electric circuit* which we are all so dependent on, and even our own *Toni's air systems*, are often disturbed by its presence. The humans around the tunnel do things that blow the fuses, send smoke through the air tubes and block our security routes. No wonder the masters of *Toni's facility management* find this tunnel a wicked problem. Still, we lifts do not forget that our predecessors were also models; we too were once made of wood and metal and glass.

Going Down – Globe

Now imagine that a scaled model wind tunnel and some people are placed in the snow globe we were introduced to when we went up. Depending on your references you will either see the activities in there around the wind tunnel as a miniature scene from the movie *Night at the Museum 1* or from Mantegna's painting *Man of Sorrows*, where the suffering Christ sits in front of a very detailed depiction of labour. In both cases the presence of the model and its workers in the globe disturbs and influences the pure swirl, whirl and twirl of the letters (or zeros and ones). When people and materials interact we observe different formations of the 'knowledge flakes'. The presence of the model attracts and structures attention anew. People at work, whether they are testing, building, looking or discussing, personify action and construct stories about what they do in the globe which we, looking through the curved surface, would want to respond to in different ways. The model as an object of engagement enables close encounters of expertise, competences and attitudes and dreams that give form to new turbulences. It is surely the place to be.

Had there been more space here an explanation of the implicit references would be in place. As for now the discussions about the agency of materials (the lift) and the necessity for a different approach to our academic data – for instance re-introducing the centrality of objects and the engagement of individuals (the globe), and a stage for performing different modes of research (wind tunnel) – are resonant in this pitch.

Taking the wind tunnel on the roof of the Zurich University of the Arts as a point of departure, I place my thoughts in local turbulences with all-embracing effects. I neither resist the pull of the depths nor the ecstasy of the heights – and so the winds embrace me, carrying me up and away.

...To the flow of people at the inauguration of the Zurich University of the Arts in the Toni Areal, an urban development project. The streamlined interest of the populace encounters an obstacle under the sky – embraces the wind tunnel, rapidly reflecting – just to enter the flow of visitors again after a single breeze.

Not only the perceiver receives of a motion
a flow of impressions,
but the world too
(even though she is not the target,
but we).

As in us,
all impressions leave firm tracks
in the world.
This is the rule:
no motion is thinkable or presentable
without her leaving traces –
be it in us,
or in the world.

A motion leaving no traces of her movement
is experienced neither by us nor by the world.
(She had better not bothered to move at all.)

A motion without track
cannot prove her movement
(for a track forms a silent witness
of any motion whatsoever).
(Raaijmakers 2009: section 186)

What remains is the amazement at the diversity of the encounters. As if the city were spitting out its people into the wind tunnel as children, only to give them back to the world again a moment later – in a kind of initiation rite – in the form of adults. But has the world been waiting for such well-organized, obliging, generous human beings? Presumably not. I will not bother all those who possess at least a hint of self-esteem with the experiences subsequently made in the maze of feasibilities and services ... lots of effort blown to the winds.

> I think it will be acknowledged as proof of the precision of my method that if one performs an experiment twice while maintaining the same conditions, the resulting images will be identical and superimposable for any point outside the area of turbulence.
> (Marey 1901: 1296, original in French)

I have personally shaped the turbulences in a project week of the research group two times. One project week was next to and on Lake Zurich and focused on the subject of models; the other was next to and around the lake and dealt with the subject of scaling. In both cases trust and confidence were produced and had an effect. Two videos reporting on these existential, confidence-building occurrences were also produced (Hothouse 1 and 2).

What began in the 19th century as a departure into the era of the fluids is still valid today: undifferentiated phenomena discover their rituals and keep circling until they arrive at specificity, where they crystallize and leave traces for the world that drifts past.

That imprints, impressions and other forms of cultural artefacts do not belong in a newly-built art college seems logical to the owner of the premises. The more time passes, however, the more a complementary approach develops out of this white-cube availability and 'bookability' – or even a genuinely urgent need for places with a character of their own. Places without a goal but with memories, strata and narratives. This motivates me to do a building project with children who perform in half-timber models, or let's call them framework models, that serve as a stage set. My research colleague fares similarly, having found this complementary approach in the guise of a former school building in the Emmental valley.

To be able to move
and move away from his place
what is moving must 'forget' his place.
But that's not all.
What is moving must forget his flight too!
(That's what fleeing is:
forgetting one's place!)

Whereas what is moving away forgets his flight,
the world manages to remember it
because she keeps track of it.
The track retains the route (the 'flight')
what is moving is taking through the world
inerasably, unchangeably and timelessly
in favour of that world.
(Raaijmakers 2009: section 187)

At the Von Karman Institute near Brussels there are loads of knowledge, skills and will when it comes to the traces left behind by objects in the wind tunnels. Indeed, while doing a tour around the VKI facilities, the subject of our conversation shifted to the memory of water as a substance and, above all, to the continuity of the inscribed trace. The examples speak for themselves: fluid intelligence thanks to an enormous power of memory. Here, to be in a state of flux is something new: remembering, forgetting – continuously!

> There is a theory which states that if ever anyone discovers exactly what the Universe is for and why it is here, it will instantly disappear and be replaced by something even more bizarre and inexplicable.
> (Adams 2005: preface)

The aerodynamic monstrosity, aka measuring device, of an era struggling to find the law of conservation of energy appears. Where do the air turbulences present a challenge to me? It's where their labels disappear, where concepts emerge that aim to represent the invisible in a visible form – in a word, between the verbal and the non-verbal. These circular air flows set projects in motion that were previously either definitely at home in the arts or definitely at home in the sciences. Thus the wind tunnel – in a circular, blowing and sucking motion – breathes linguistic and non-linguistic phenomena from one world into the other. At a university, I localize the turbulences that initiate a maturation process in the so-called third cycle – that place where, otherwise,

References

Adams, Douglas, 2005, *The Restaurant at the End of the Universe*. New York: Ballantine Books.

Marey, Étienne-Jules, 1901, 'Changements de direction et de vitesse d'un courant d'air qui rencontre des corps de formes diverses' in *Comptes rendus hebdomadaires des séances de l'Académie des Sciences* 132: 1291–1296.

Raaijmakers [Raaymakers], Dick, 2009, Method (divided by sections; ed. and trans. by Vincent W.J. van Gerven Oei). Eindhoven: Onomatopee. Originally published in Dutch as *De methode*. Amsterdam: Bert Bakker, 1985.

Serres, Michel, 1977, *Hermès IV: La distribution*. Paris: Les Éditions de Minuit.

artistic doctoral dissertations putter about. Dispensing with theses, I place the world in straightened air and observe its turbulences. They tell pre-specific narratives waiting for their adaptation.

...Quite like in those times when systems became fluid. The turbulences remain.

> In the beginning there is the undifferentiable, about which nobody can possess information. This can be called a cloud. A set of points, of atoms or molecules, of entirely arbitrary elements, whose behaviour is unknown, clouds with undefined, fluctuating, or blurring edges. [...] This meteorological phenomenon is the model of a knowledge formed shortly before the turn of the century, when the systems faded [...]. (Serres 1977: 9–10, original in French)

English translation by Sebastian Viebahn

Of Cities and Pinpoints; or, Where Do Literature and Wind Tunnels Meet?

Mirjam Steiner

Taking the wind tunnel as a point of departure, I reflect upon how sizes and scales are whirled around and disrupted in literary texts – something that is never harmless. For texts that engage with the concepts of large and small are full of absurdity, paradox, misunderstanding, exaggeration and alienation.

Put to the Test

Wind tunnels are used to test, model and verify. They are concerned with speeds, forces and wind loads. Objects commonly investigated include airplanes, cars and trains, but models of entire cities are also measured in wind tunnels. The stated goal is a reality check based upon a relation of similarity: the scaled-down models are calibrated using real data and their validity for the full-scale object is tested. In the process, laminar flows alternate with turbulences.

Test situations with scaled figures or objects also occur in literary texts. Alice in Wonderland repeatedly changes size as soon as she eats or drinks anything; on the one hand, this tests her ability to deal with her shifting size, while on the other hand communication is also subjected to close scrutiny. Alice has arrived in a country full of absurdity and paradox, where animals and playing cards speak and cats grin and vanish. Every conversation between Alice and the bizarre characters of Wonderland is doomed to failure. She is constantly rebuked and confronted with misunderstanding, and her (school) knowledge – which she makes repeated attempts to verify – proves to be inadequate. Time and space begin to totter; all familiar principles of order are suspended. The world of Wonderland is determined by nonsense. Alice's changing body size follows no rational logic and has its counterpart in Carroll's representation of the fragility of language.

Turbulences

Aerodynamics has taught us that scaling effects sometimes lead to a number of complications. Successfully testing the model of an airplane in a wind tunnel by no means guarantees that the prototype will be able to fly, as the model's aerodynamic qualities

cannot be transferred automatically. Physics makes use of the Reynolds number among others in order to bring original and model closer together.

In literature, scaling effects are not without consequences either, as the nameless protagonist of Will Self's story *Scale* is forced to learn. "Some people lose their sense of proportion; I've lost my sense of scale" (Self 1995: 3), he observes at the beginning of the story, and as the narrative progresses we gradually realize that he has lost his sense not only of scale, but also of any moderation and is moving through a gradually shrinking world in a state of drug-induced intoxication. During a nocturnal drug trip, the first person narrator blunders ever deeper into a world full of model constructions: he moves from a model village to a model house, then to the next smaller model house and so on. Analogously to the constantly dwindling spaces, the protagonist himself shrinks continuously until he is as small as a subatomic particle, fearing that he will cease to exist as an entity on the submicroscopic level. To his eyes, wasps appear like zeppelins and drifting dust bunnies like thunderous avalanches, confounding familiar orders of magnitude. While the size of the protagonist's body decreases in proportion with the shrinking model village surrounding him, his sense of time does not behave analogously. To the contrary: his return seems to him to take several months. Nor does his trip begin to fade, but rather manifests itself in a general loss of a sense of reality that goes hand in hand with a feeling of insecurity leading to existential crisis.

In *Scale*, scaling effects present the protagonist's loss of identity; in *Gulliver's Travels*, however, they are used as a device to illustrate the relativity of all insight. Unlike with Alice or the protagonist of *Scale*, the size of Gulliver's body does not change. Depending on the region of the world he is travelling through, the human beings surrounding him are sometimes small, sometimes large. While the Lilliputians fear Gulliver, the man-mountain, for quite a long time he idealizes the miniature world that he is able to survey from a bird's-eye perspective. The appropriateness of proportions within one's respective community fails to function when comparing oneself to the size of the other. Thus the gigantic Gulliver is no longer able to perceive the sewing Lilliputian girl's thread and needle with his naked eye. From his point of view, the world of Lilliput is microscopic; things that seem familiar are rendered unrecognizable due to their reduction in size. At the same time, the differences in size collapse here, too, when Gulliver extinguishes the fire that has broken out in the imperial palace with a gigantic stream of urine, of all things – for which he is branded a traitor.

References

Self, Will, 1995, *Scale*. London: Penguin.

Grotesque events go hand in hand with scaling effects when Gulliver is exposed to all kinds of monstrosities in Brobdingnag, where he mutates into a dwarf among giants. Close-ups of the monstrously large, which sometimes can be an infant, sometimes a naked breast, lead Gulliver to experience human beings and body parts that he would find attractive under normal circumstances as threatening and repulsive. Under his gaze, which captures only those details of the giant world closest to him, proportions and symmetries shift, harmony is disrupted, and beauty and ugliness become a matter of perspective. As a miniature human being among giants, he experiences reality as exaggerated, deformed and distorted.

Becoming visible

In the repertory of literary motifs and characters, scaling is used as a device to present polar structures. Scaling achieves its greatest effect when large and small are set side by side. Only the juxtaposition of two disparate sizes clearly reveals the corresponding traits: the absurd, the grotesque and the nonsensical only emerge when contrasted with a dominant order (of magnitude) that functions as a point of reference. Accordingly, large and small are not only relative, but inextricably connected.

Literary texts that present scaling effects possess discursive potential: they test values, knowledge and hierarchies. And – analogously to the wind tunnel – they function as an aesthetic experimental set-up in which various stresses and loads can be investigated and negotiated before they occur in reality.

In the same way that flow phenomena are visualized in wind tunnels using threads, smoke and fog, producing strangely beautiful images, literary texts have the potential to create images before the reader's inner eye by transforming large and small. Both what is visible and what is concealed are revealed by the driving forces of the wind and the imagination.

English translation by Margaret Hiley

Ma[1]
Florian Dombois

Taking the wind tunnel as a point of departure, I start to think: how are we to imagine a 'research focus on transdisciplinarity' at art colleges? What could 'art as research' mean, and how should this double aspiration be permitted to articulate itself so that the provocation arising from this oxymoron can work in both directions, leading to real movement of disciplines?

What happens to art when it claims referentiality in its works and even wants this referentiality to be discussed using language? Does this take us back to the 19th century, exposing us to the threat of an aesthetic morality squad, or is there a way that non-verbal formulation, political commitment and conceptual clarity can be brought together? What kind of aesthetic thought would make verbal demands to then take risks of its own in a non-verbal manner? What would art that claims to be as poetic on the conceptual level as on the material look like?

And vice versa, how would scientific research develop if its focus were no longer to solve problems, if 'knowledge' were no longer dealt with as a reproducible commodity, but 'science' – the creation of knowledge – were taken seriously as an epistemic practice in its temporality and processuality? And what would happen if scientific research were to develop the scope of the media it uses to articulate itself beyond language? How would it engage with those ideas that can only be formulated in a non-verbal and non-propositional manner?

I turn around and look back at the test section of our wind tunnel. On the right-hand side I can see the nozzle from which the pressurised, accelerated air emerges to then disappear, sucked into the opening of the funnel. Or is it the other way round, and the wind emerges from the suction tube, reaching into the nozzle from there?

If I introduce a cloud of smoke into this air current, it reappears after a certain time. Then it has passed once through the entire length of the wind tunnel. And I ask myself: is it still possible to create art at an art college today? What are the conditions of the institution to which students and teachers are subject in equal

measure? What does the institution do with us, what does it do with our work? And vice versa, what actions are the institution's internal dynamics responding to, where do they escalate?

The smoke alarm above the wind tunnel needs to be turned off each time we conduct an experiment. We notify our facility management of this a day in advance at the latest. Our wind tunnel is on the roof. The real wind is often stronger than the wind generated by the fan. It whirls and messes up the measuring section. An escape route runs through our laboratory. Facility management says it cannot be re-routed. In the mornings, we find cigarette butts in our laboratory. We are public. At night, too, others come here.

How must a 'research focus on transdisciplinarity' be set up in order to enable different disciplines to meet as equals? How can their perspectives be bundled? How to provide everyone who comes with the space that they need? Where and how can they show what they think? What they know? What they ask themselves and others? Is there a stage on which anything – anything at all – can be performed? And what happens to those watching this stage, which no longer knows any clearly defined disciplines, where one never knows whether one is witnessing a drama, a lecture or a stand-up comedy? Is this even bearable?

When first visiting our wind tunnel, Hans Jörg Rheinberger called it "a model of a second order and at the same time unsettlingly specific". Does the opposite apply, too? When do models become unsettling? And what does the mind do when constantly scaling to and fro between the concrete and the metaphorical? Can this oscillatory movement be called aesthetic? Pavel Büchler said during his first visit: "Is it art? What else could it be!" Perhaps we need to search for art where it no longer claims to be art.

Scientific research is part of a food chain: it supplies our industry – this is at least the way current politics would have it. Can art also claim to be supplied via such a food chain? In other words, can art claim that research should be conducted at art college in order to help art to supply the art market? What do the artists on the art market actually need in order to produce good work? What would 'research findings' with the potential to be used further in artistic production look like? And might there even be findings that could also be used in scientific production? What would this kind of hybrid beings look like?

I love the wind. You can never see the wind itself, only its effects. You can never hear it, only its effects. It needs others to reveal itself, and thus it is as if made for transdisciplinarity. It

References

Groys, Boris, 1997, 'Das Museum im Zeitalter der Medien' in *Logik der Sammlung: Am Ende des musealen Zeitalters*. München and Wien: Hanser: 7–24.

can be gentle as a breeze or strong as a gale. First it refreshes, then it terrifies you. It is so rich, so much at once. It acts as an equalizer between areas of high and low pressure. Chinese medicine imagined it as flowing from the heart to the kidneys of healthy human beings. Is the nozzle the heart and the suction tube the kidney?

For Aristotle, winds explained earthquakes: when they entered the underworld, the earth shook. With the wind, the sky reached into the earth. Aristotle developed his explanation of earthquakes in his book on meteorology and thus placed these seismic phenomena seamlessly between a chapter about the wind rose and one about lightning, thunder and whirlwinds. The Greek word μετέωρος means to 'hang in the air', but also 'changeable, uncertain'. Even today, seismology and meteorology are often found in the same university faculty.

We speak of wind as a thing. Martin Burr once brought up the question: where is the wind when it is not blowing? Research should be posed this question! Does knowledge even exist if nobody is working on it? Where is science when it is not researching? Where is society when it is not living? And where is art when it is not acting? Boris Groys once described museums as "graveyards of things" (Groys 1997: 8, original in German). There are many graveyards of knowledge. And they are well paid.

Swiss polytechnics have a political mission to carry out research. Swiss colleges of art (also called *universities of the arts*) are polytechnics. Bringing academic research to the art colleges and carrying out research projects with its help is easy. This challenges neither the arts nor the sciences. But I refuse to take the easy way out. I have not yet abandoned my desire to attempt a researching art practice at the art colleges that serves the disciplines taught there. Something that drives the arts themselves forward. Something that both emerging and established artists can draw upon.

English translation by Margaret Hiley

1 *Ma* is a Japanese concept describing the negative space, the gap, the pause between things. It treats something as a thing that we are used to seeing as no-thing. *Ma* brings the background to the fore, the boundary conditions to the centre. It makes us aware of what we are used to overlooking. This is what we are working with. Our wind tunnel encloses a free space in which thoughts can move.

The organizational structure of the project 'Size Matters' was based on the concept that the researchers and other participants, but mainly the project's team, would gather regularly and interact around different and differentiated tests involving the central object or instrument, the wind tunnel. The team's exchanges alternated between working in non-verbal or verbal, practice- or language-based modes, which could be spontaneous or meticulously planned. Having each contributed to the immediate interaction in our own way, we would then integrate narratives and conclusions from the collaborative research into our individual work. At the centre of this there was a common thread of images, films and photos, texts and objects created in the process and diligently collected, organized and preserved. On the following pages the project's 36 months provide a basic (if abstract) sequence, which enables us to focus on some of the themes that were central throughout the funded period.

For each of the months you will find: i) a short text, ii) a diary of events and iii) images. The diary (ii) has been split into the following categories: *Meeting*, *Construction*, *Publication*, *Multiple*, *Excursion*, *Teaching*, *Conversation*, *Hothouse*, *Bulletin*, *Artist/Writer in Residence*, *Modellapéro* and *Team*. These 12 categories highlight modes of interaction between the team, students, professionals and various other forms of audiences. Some names and quotes appear, but they must be seen as representative of many more individuals who, in different ways, were part of the wind tunnel's open studio-laboratory. The accompanying texts reflect how the exchanges were most often in-between; they lean into the realms of science, art and history, but do not adhere to one particular form of knowledge or discipline. Dates are indicated in the format YY-MM-DD. The team members were Florian Dombois FD, Haseeb Ahmed HA, Martin Burr MB, Julie Harboe JH (July 2014–December 2015), Kaspar König KK, Mirjam Steiner MS, Sarine Waltenspül SW and Reinhard Wendler RW (January 2013–June 2014).

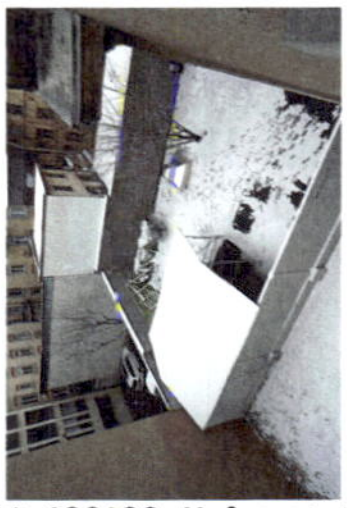
***_130100_Hafnerstr.tif

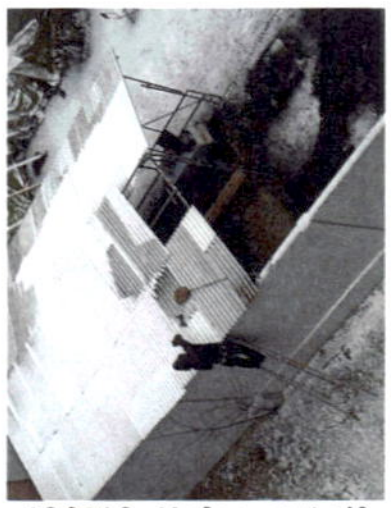
130116_Hafnerstr1.tif

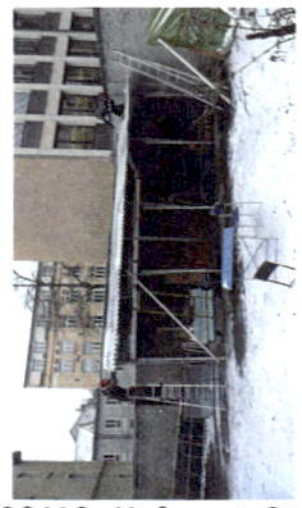
130116_Hafnerstr2.tif

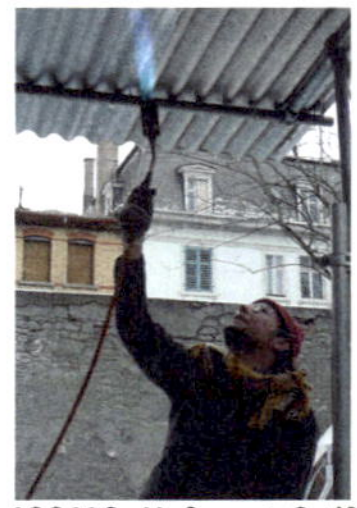
130116_Hafnerstr3.tif

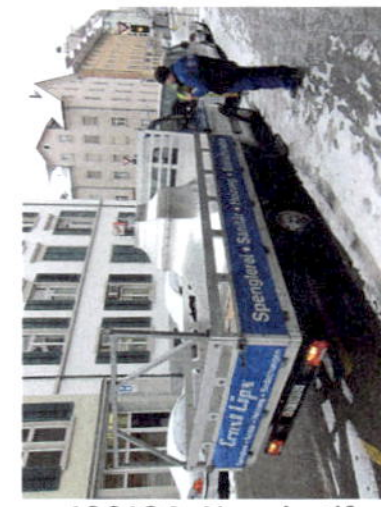
130124_Nozzle.tif

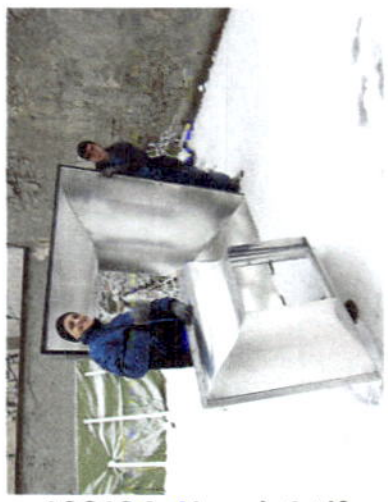
130124_Nozzle1.tif

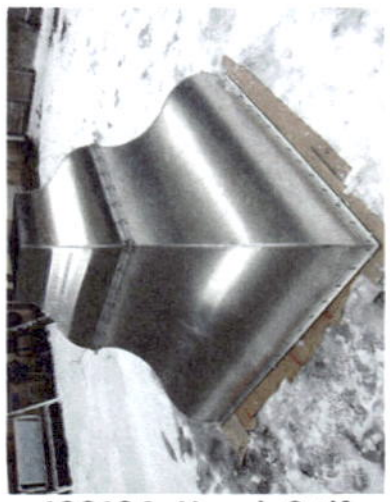
130124_Nozzle2.tif

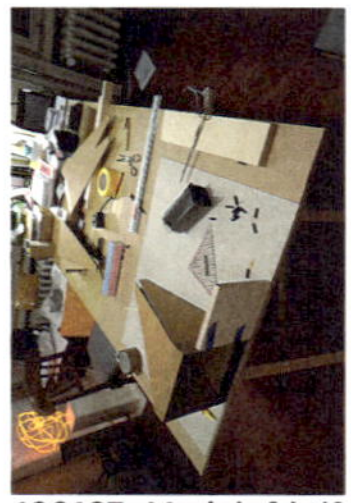
130127_Model_fd.tif

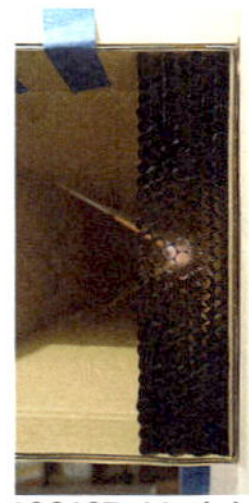
130127_Model_fd1.tif

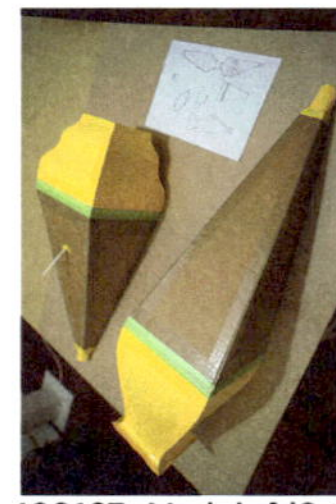
130127_Model_fd2.tif

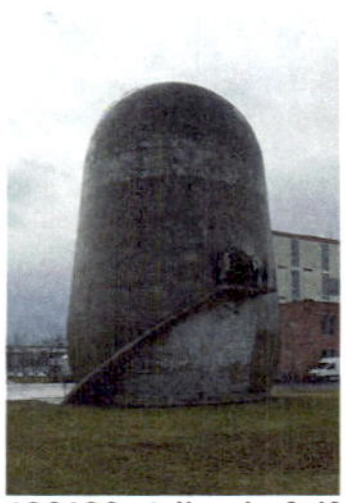
130130_Adlershof.tif

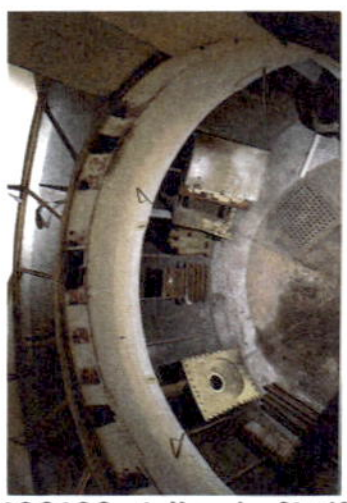
130130_Adlershof1.tif

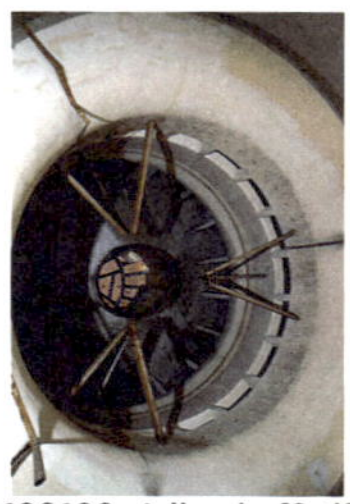
130130_Adlershof2.tif

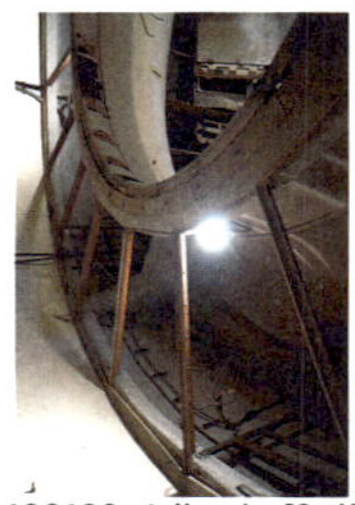
130130_Adlershof3.tif

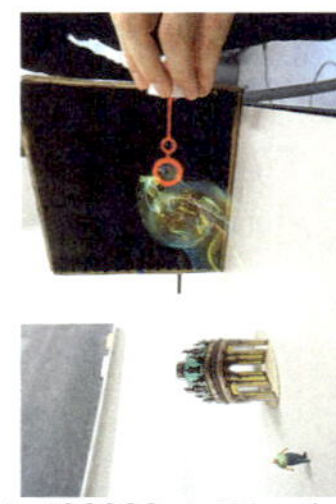
***_130200_1stMeeting.ti

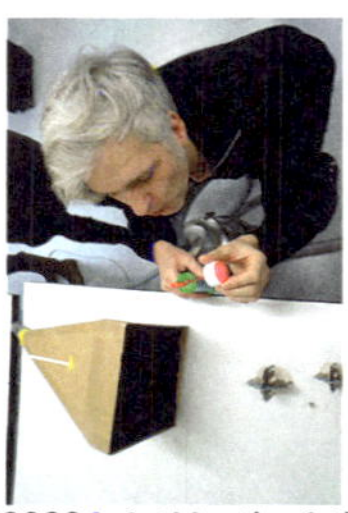
130204_1stMeeting1.tif

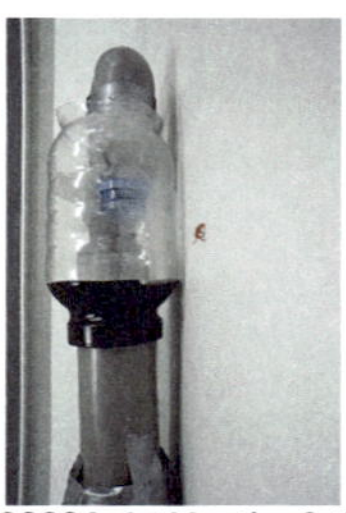
130204_1stMeeting2.tif

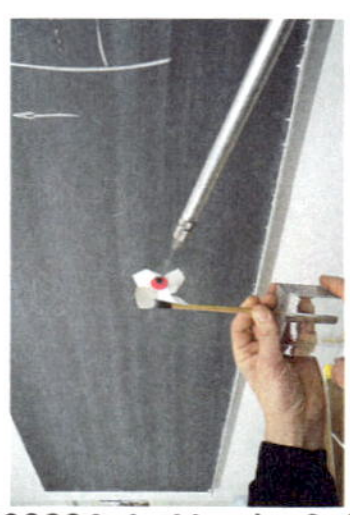
130204_1stMeeting3.tif

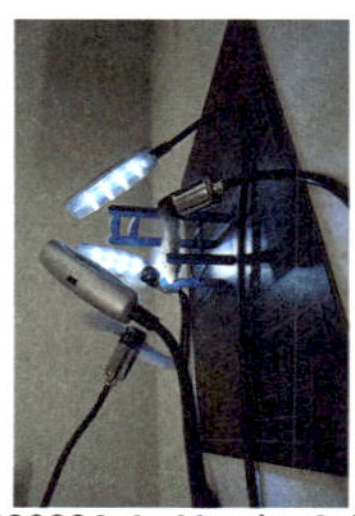
130204_1stMeeting4.tif

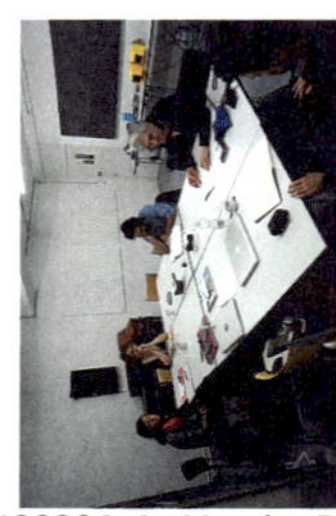
130204_1stMeeting5.tif

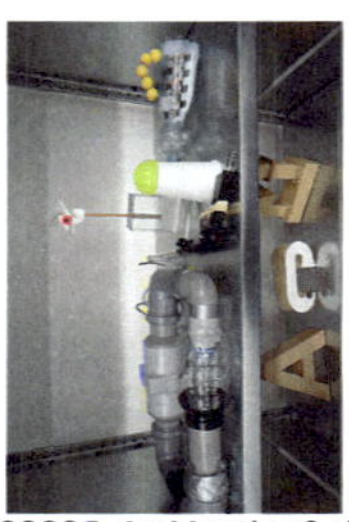
130205_1stMeeting6.tif

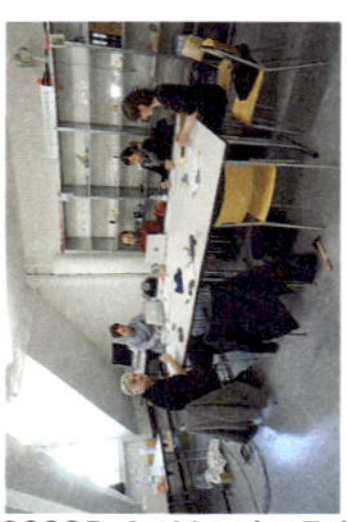
130205_1stMeeting7.tif

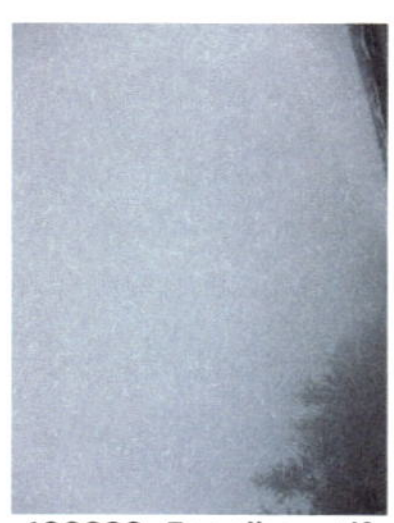
130220_Engelberg.tif

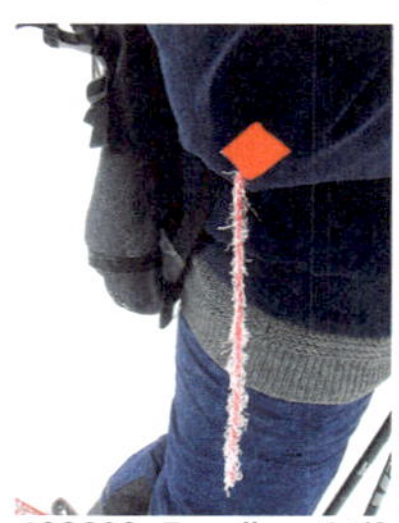
130220_Engelberg1.tif

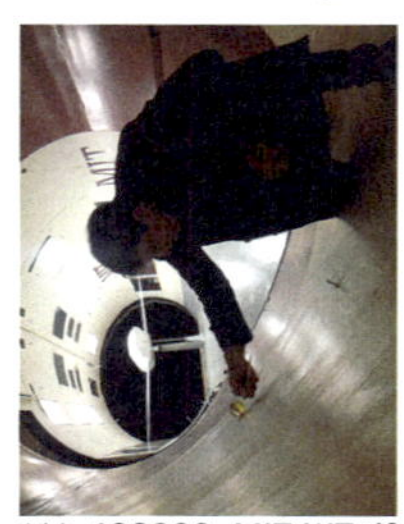
***_130300_MIT-WT.tif

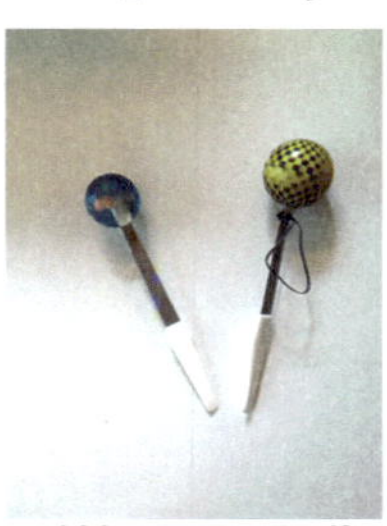
130311_MIT-WT1.tif

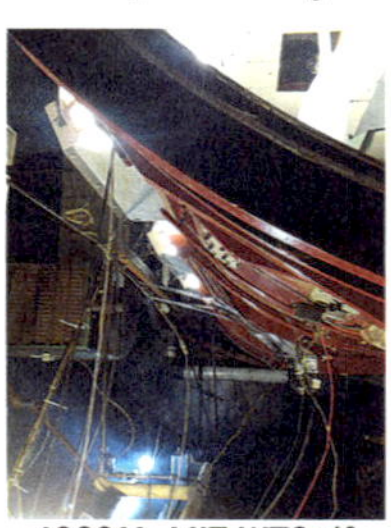
130311_MIT-WT2.tif

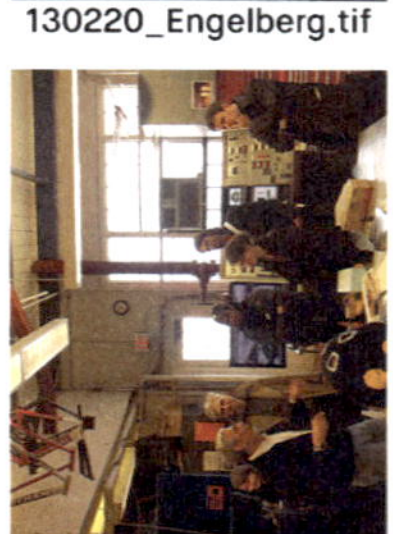
130311_MIT-WT3.tif

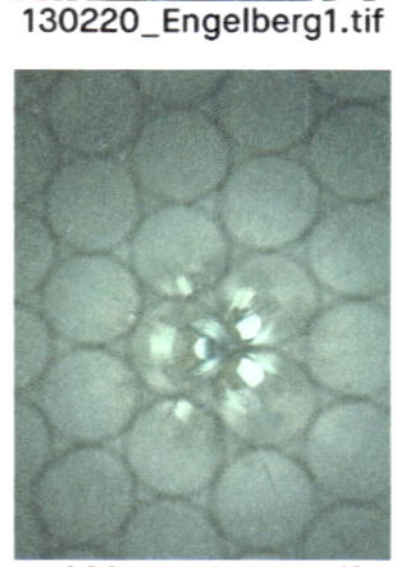
130311_MIT-WT4.tif

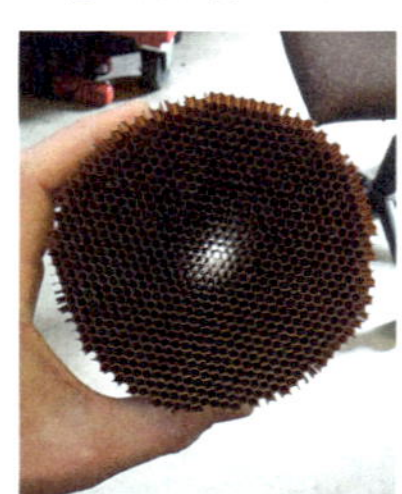
130315_MIT-honeyc.tif

Jan 2013
Nozzle

The planning of the wind tunnel started in the summer of 2012. It was based on sketches and discussions between FD and KK. By autumn of the same year, the first digital 3D models had been developed. The wind tunnel was to be assembled in the yard of the Zurich University of the Arts building in the Hafnerstrasse, close to the city centre. It was constructed out of modular elements, which would enable it to be easily changed, expanded and moved. The nozzle was a key element of the apparatus, and it arrived in January of 2013 after specialists had finished making it. As the nozzle narrows, the wind is condensed and accelerated.

130101
The official start of the SNSF project 'Size Matters: On Size and Scale of Models'.

130116
Construction
Roof of the wind tunnel shed is built at Hafnerstrasse.

130124
Construction
Nozzle arrives.

130130
Excursion
Team meets for the first time in Berlin and visits the Teaching Model Collection of Humboldt University and the Aerodynamic Park Adlershof (see May 2014).

Feb 2013
Meetings

Throughout the project the 'Size Matters' team met regularly for workshops. Each meeting was dedicated to a theme. The approach followed the concept of *Fahrkunst* (see March 2015), where discussion and exercises alternate between the verbal and non-verbal and between practical and theoretical work. This reflected the diverse expertise in the team: artists and cultural historians. At the first meeting everybody received the assignment: "Build a functioning model-size wind tunnel." This led to interpretations of the theme ranging from the very abstract narrative of a bottle-shaped tunnel as a Noah's Ark featuring miniature animals to more traditional variations on scientific wind tunnels.

130204–130205
Meeting
First of the monthly group meetings.

130220
Excursion
Skiing in Engelberg. The weather conditions reduce visibility to only two metres.

130221
Teaching
FD and RW start teaching on *Model Theories in the Wind Tunnel* in the MA Transdisciplinary Studies programme (see March 2013, April 2013).

Mar 2013
MIT Wind Tunnel

In March the team and students visited Boston and the Wright Brothers Wind Tunnel at the Massachusetts Institute of Technology (MIT). The aeronautic pioneer Orville Wright had attended the tunnel's inauguration in 1938. No other laboratory on MIT's campus has survived for so long. After 75 years it is still used by students for instruction and research and it is largely unchanged. At different times HA, FD and Luis Berríos-Negrón (see Artist in Residence, October and December 2013) have all found inspiration and conducted artistic work in this particular tunnel.

130306–130308
Construction
KK sews the laboratory tent. It is a white cube around the tunnel, which protects it from wind and rain.

130309–130315
Excursion
Team takes the master students to Boston, visiting e.g. the Wright Brothers Wind Tunnel at MIT and the Glass Flower Collection at Harvard, and it conducts a workshop on models with staff of the MIT Museum.

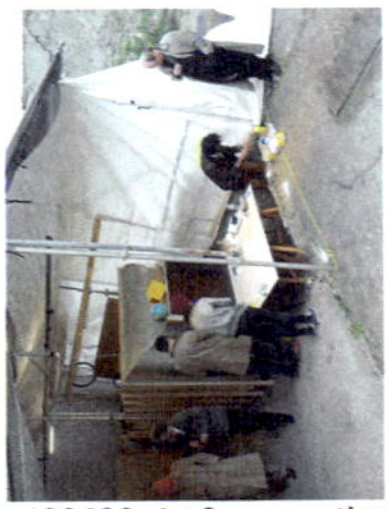
***_130400_1stConversation.tif
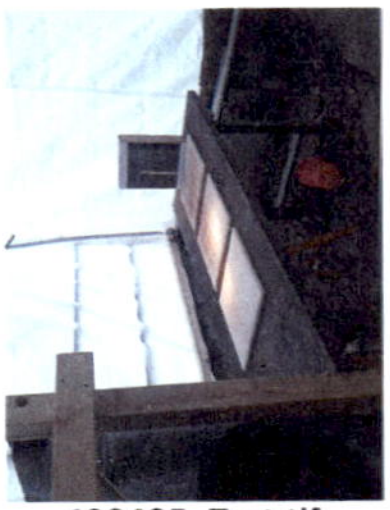
130405_Test.tif
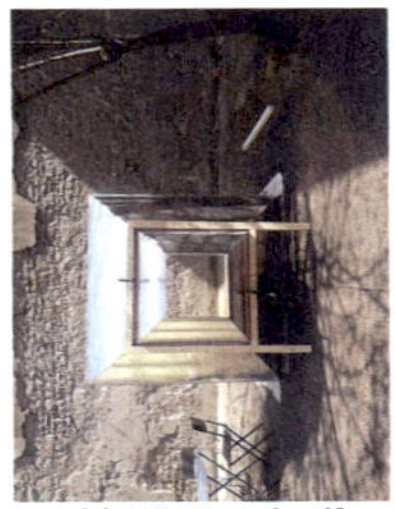
130417_Nozzle.tif
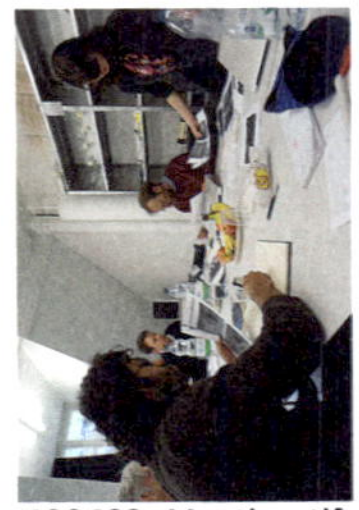
130429_Meeting.tif
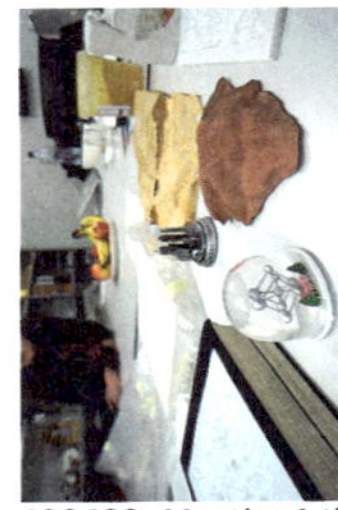
130429_Meeting1.tif
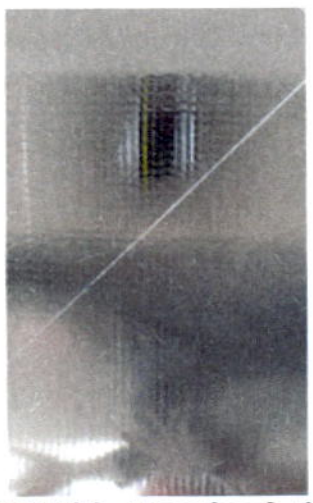
130429_Meeting2.tif
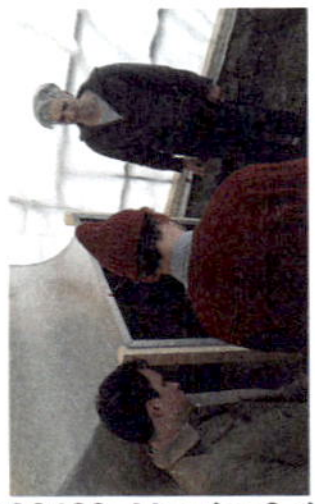
130429_Meeting3.tif
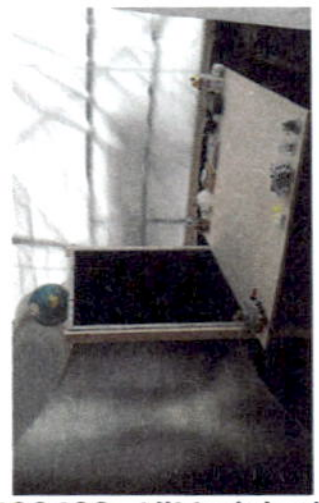
130430_AllModels.tif
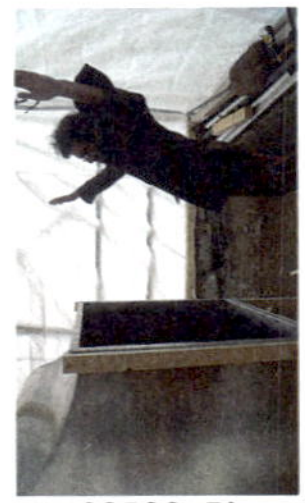
***_130500_Blow.tif
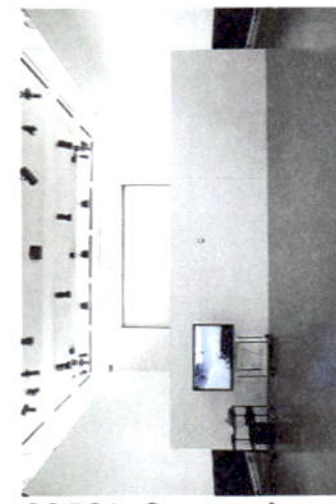
130521_Symposium.tif
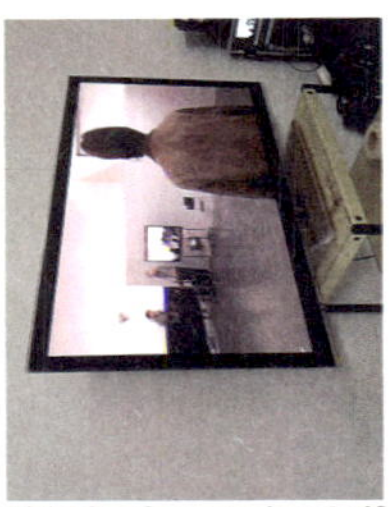
130521_Symposium1.tif
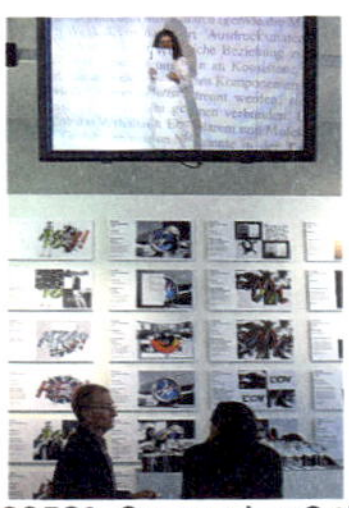
130521_Symposium2.tif
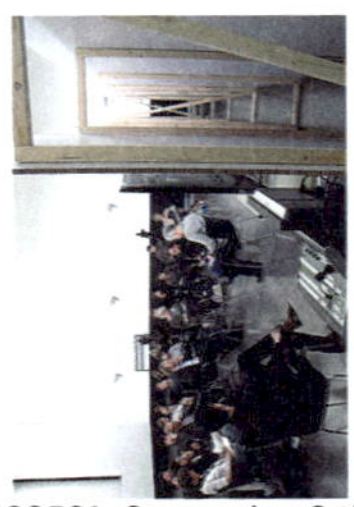
130521_Symposium3.tif
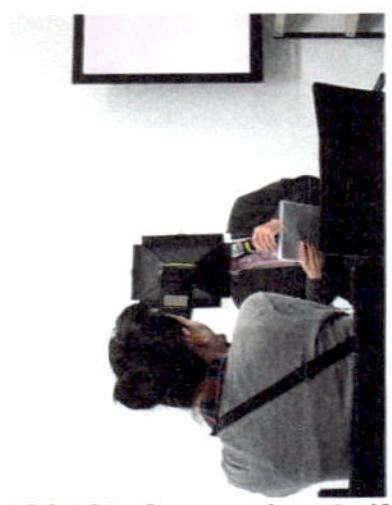
130521_Symposium4.tif
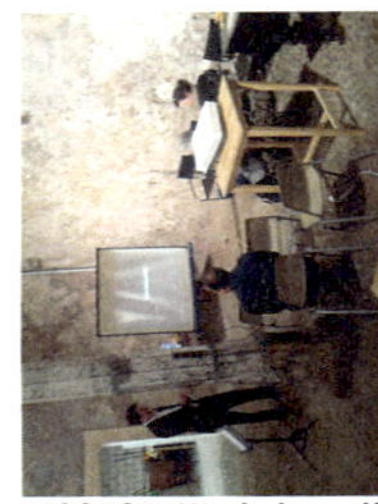
130524_Workshop.tif
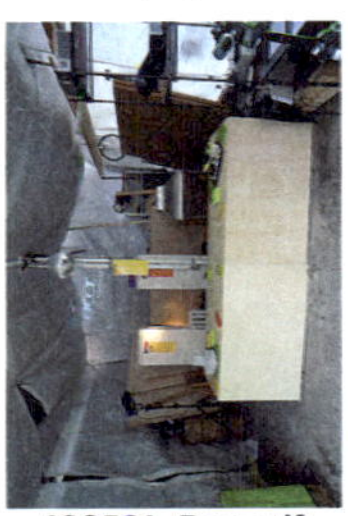
130531_Party.tif
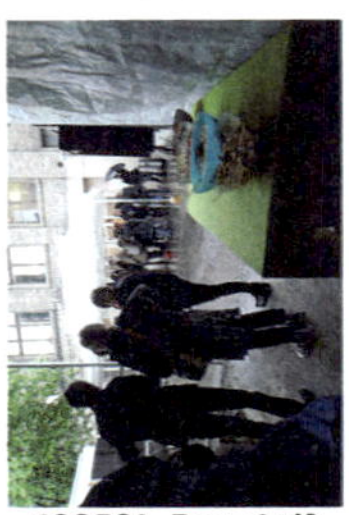
130531_Party1.tif
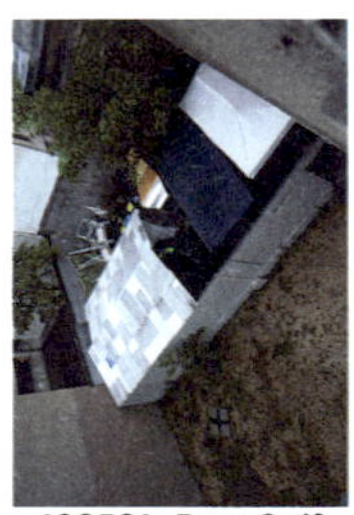
130531_Party2.tif
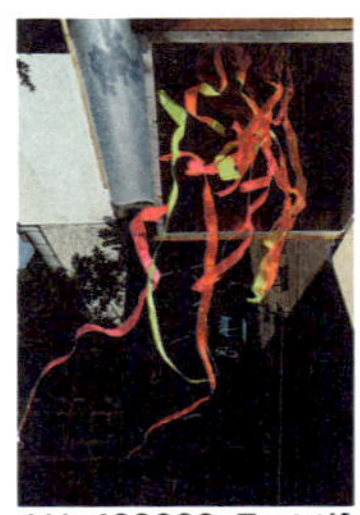
***_130600_Test.tif
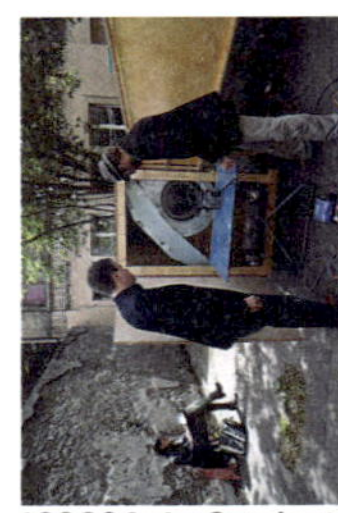
130604_1stSmoke.tif
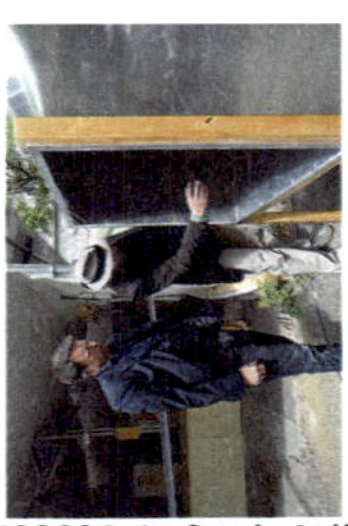
130604_1stSmoke1.tif
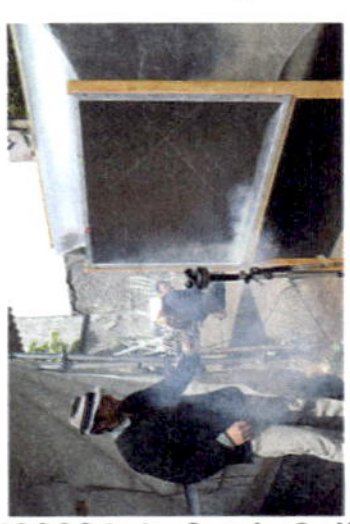
130604_1stSmoke2.tif
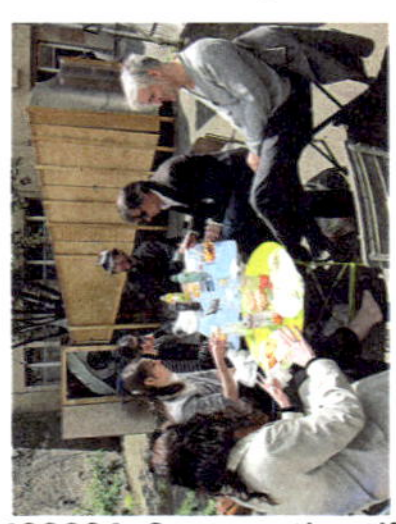
130604_Conversation.tif
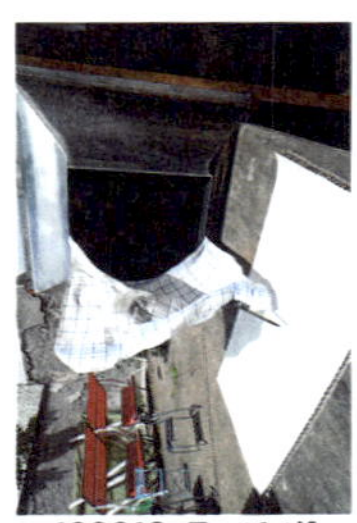
130612_Test1.tif
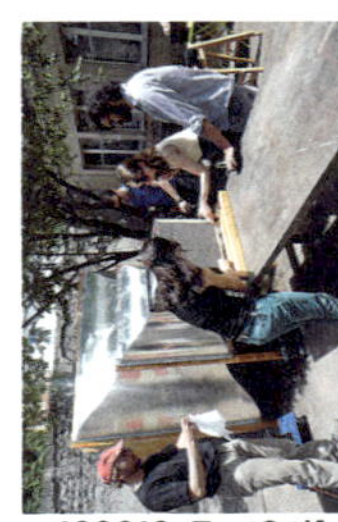
130612_Test2.tif
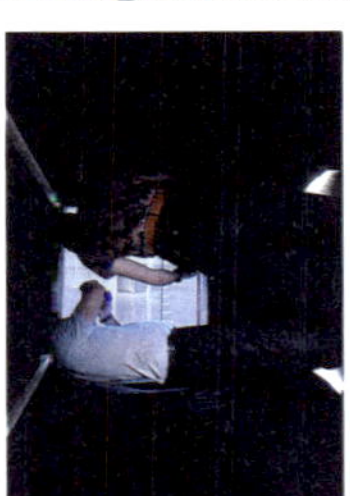
130613_Test3.tif
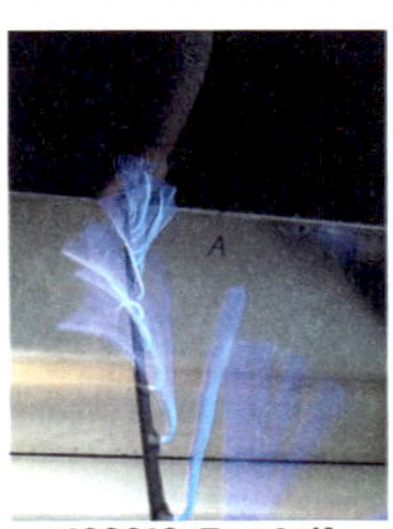
130613_Test4.tif
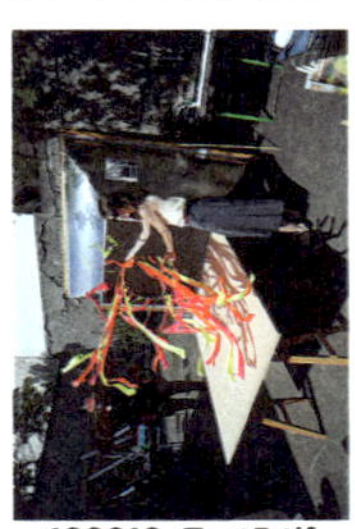
130613_Test5.tif
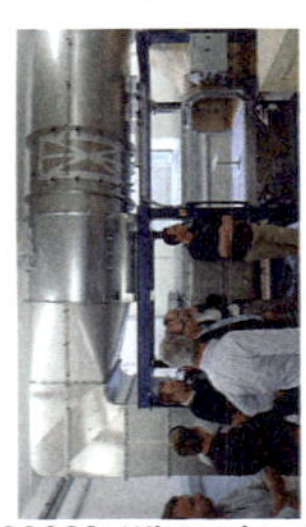
130628_Winterthur.tif
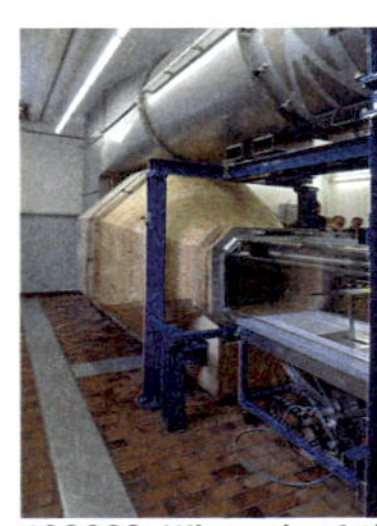
130628_Winterthur1.tif

Apr 2013
Conversations

'The Wind Tunnel Conversation' was developed as a format for intellectual and artistic discussions between the members of the team and invited guests. Scientists, laypersons, artists and designers met and engaged in a direct form of transdisciplinary exchange. The conversations had no fixed schedule and did not adhere to traditional academic forms of presentation. Instead, the guests visited the wind tunnel and acted with it. The proximity of this ambiguous instrument opened up the possibility for discussions on different kinds of matters.

130410
Construction
The engine is equipped with a new transmission belt and continuous control unit.

130421–130422
Hothouse
In preparation for the Hothouse in the autumn, the team produces the booklet *Scale Model Wind Emblemata* as a gift to the three specially invited guests.

130429
Construction
The first four modules of the wind tunnel are finished (engine, duct segment with expanding cross section, straightener, nozzle) and combined, so that the wind blows through the nozzle.

130429–130430
Meeting
Collective readings about scale.

130430
Conversation
Bernd Mahr, model theoretician and mathematician from the TU Berlin, and Hans-Jörg Rheinberger, director of the Max-Planck-Institute for the History of Science Berlin, are the first guests in the open series of wind tunnel conversations.

Hans-Jörg Rheinberger:
The wind tunnel is a model of a second order and at the same time unsettlingly specific.

May 2013
Tunnel Attractions

Sensory impressions play an important part in the research surrounding the tunnel. Standing in the wind and thinking in the wind has an impact on both body and mind. Today much research into fluid dynamics is conducted with computer modelling. The wind tunnel community (see September 2015) is exceptional in that it relies on material evidence as its point of departure. This demands training and a trust in tacit knowledge. Sometimes it seemed that the alchemic traditions, essential to the proto-scientific experience and not unfamiliar to the arts, could be awakened by the 'Temple of the Winds'.

130508
Construction
Work begins on the mould for the expanded curve (see July 2013)

130514
Publication
FD gives a lecture at the Artistic PhD Programme of St Lucas University College of Art & Design in Antwerp. This will eventually be the host institution for HA's PhD.

130521
Publication
The symposium *Nicht-Propositionalität und ästhetisches Denken* (Non-propositionality and aesthetic thinking) makes use of the scaling set-up of 'Palaver' (see May 2015 and Selected Publications).

Jun 2013
Visualizing Wind

Through ribbons, cloths and threads the interaction between the wind and the objects waved their visibility at their viewers. In the June meeting, the team generated tests that displayed movements: colourful and nuanced, wild and delicate. Each test was recorded and its title, author and date were noted. A series of videos were made accessible on the first public online archive for artistic research, the Research Catalogue (see https://www.researchcatalogue.net/view/56777/56778). Throughout the project this video log would provide some order and insight into those activities that could only be captured by the moving image.

130604
Conversation
Juliet Koss, art historian and model researcher from New York and Los Angeles, visits the wind tunnel. The subsequent discussion covers themes such as cameras, models and their use in the early Soviet avant-garde. Johannes Fehr, vice-director of Collegium Helveticum also joins the wind tunnel conversation.

130606
Teaching
Students from the ZHdK design department test their model airplanes in the tunnel. This is the beginning of non-scheduled use of the wind tunnel by ZHdK students.

130612–130613
Meeting
Visualizing techniques and conceptual work on a new proto-disciplinary publication format that later becomes the *Wind Tunnel Bulletin* (see November 2013, April 2014, November 2014, June 2015).

Juliet Koss:
That wind tunnel is like an adolescent. You produced him and now he's living in your house, taking up all this space, getting bigger and bigger.
It's not clear what he wants, but he definitely wants something, and he's a pain in the ass.

Johannes Fehr:
The answer is blowing in the machine.

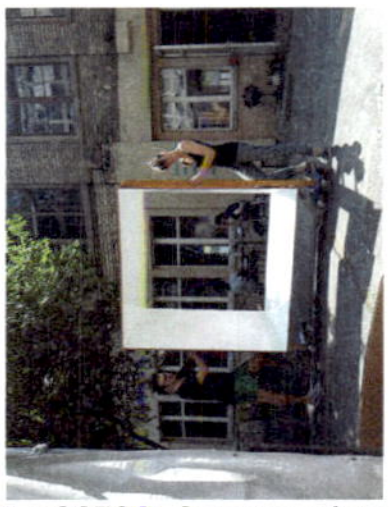
***_130700_Construction.tif

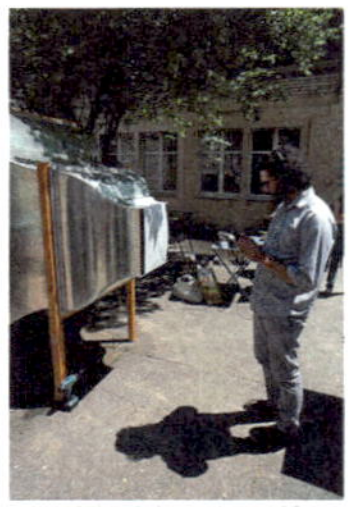
130702_Test.tif

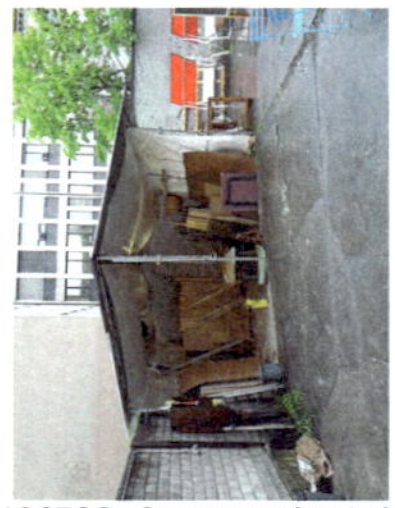
130703_Construction1.tif

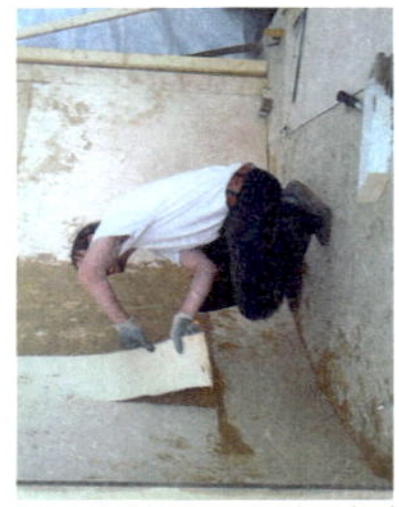
130704_Construction2.tif

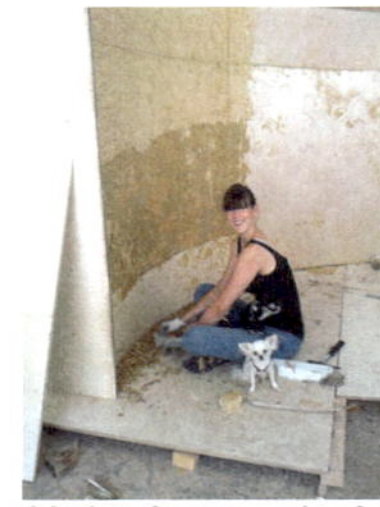
130704_Construction3.tif

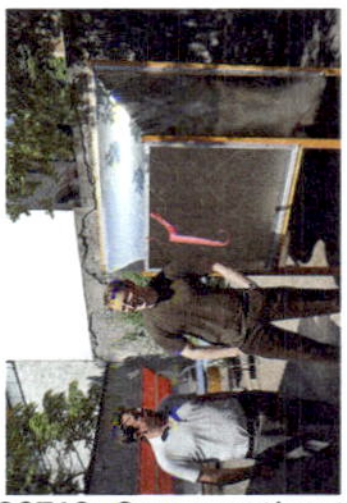
130710_Conversation.tif

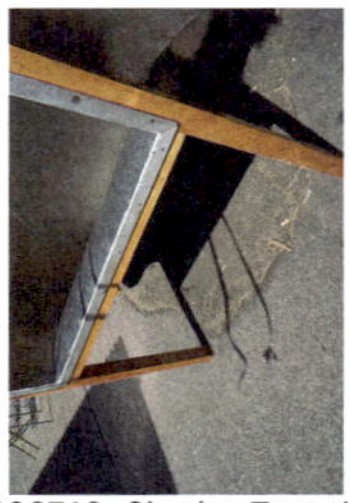
130710_ShadowTest.tif

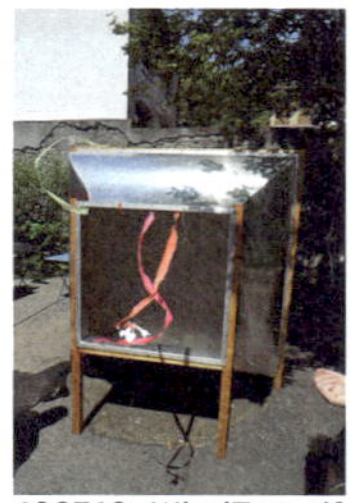
130710_WindTest.tif

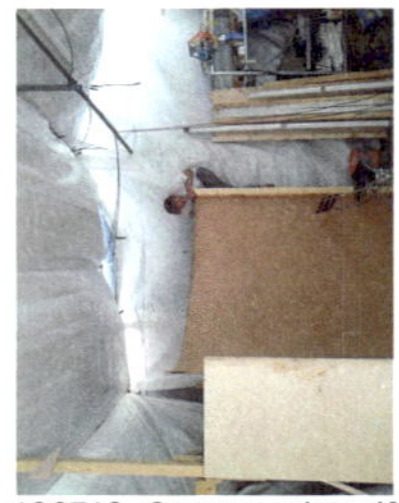
130718_Construction.tif

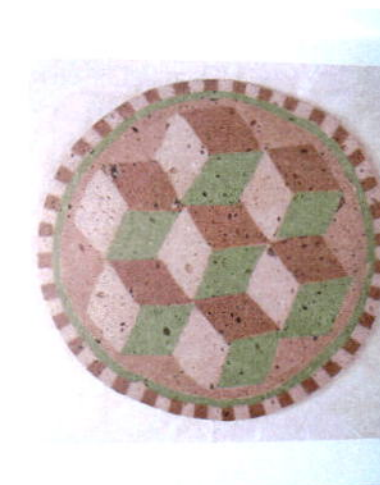
130722_TestBologna.tif

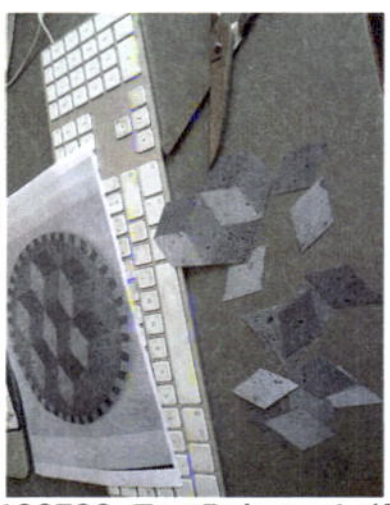
130722_TestBologna1.tif

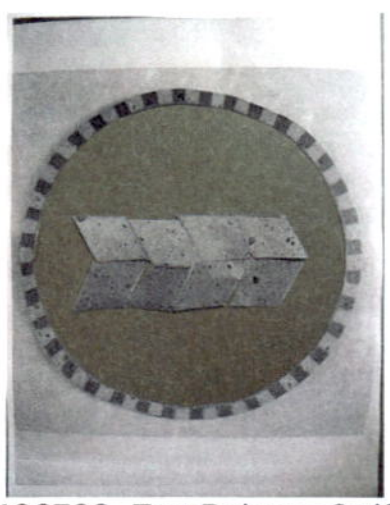
130722_TestBologna2.tif

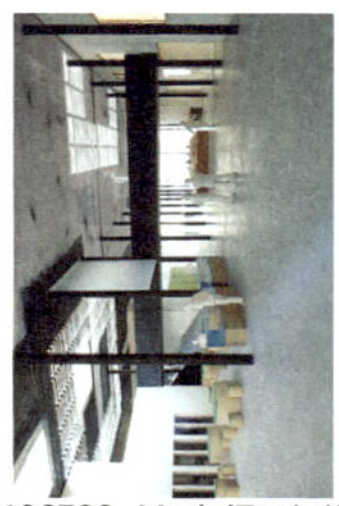
130723_ModelToni.tif

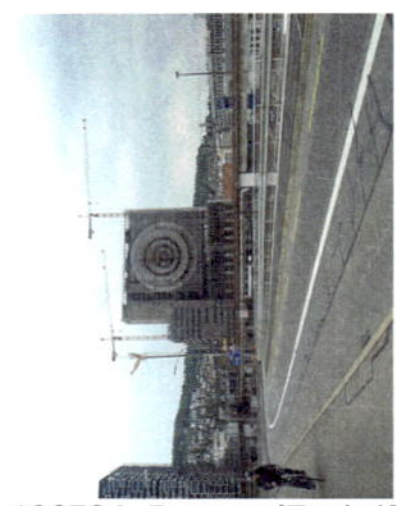
130724_ProposalToni.tif

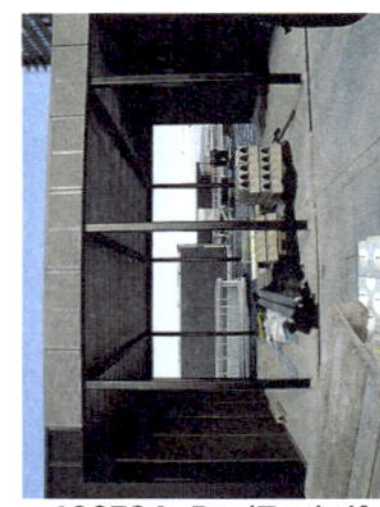
130724_RealToni.tif

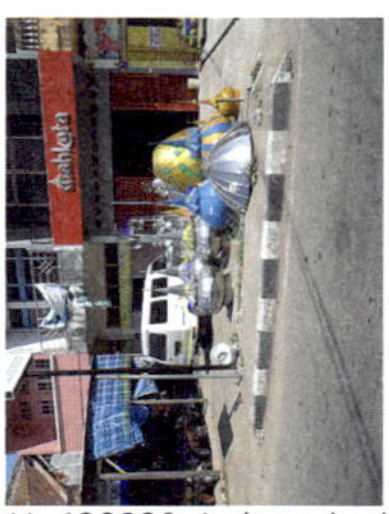
***_130800_Indonesia.tif

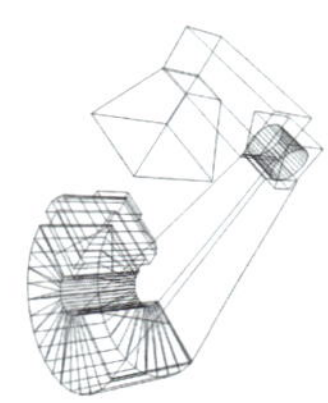
130822_Wireframe.tif

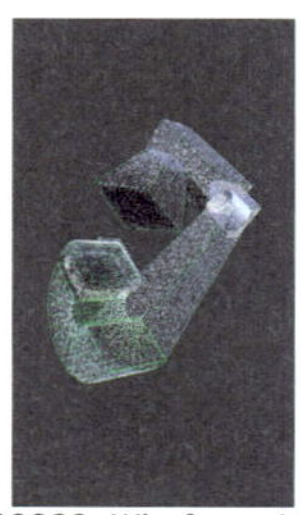
130822_Wireframe1.tif

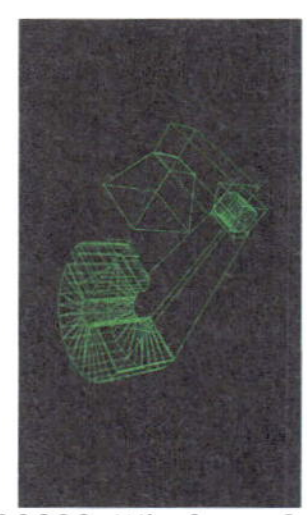
130822_Wireframe2.tif

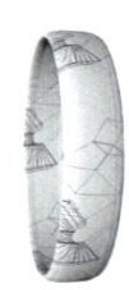
130822_Wireframe3.tif

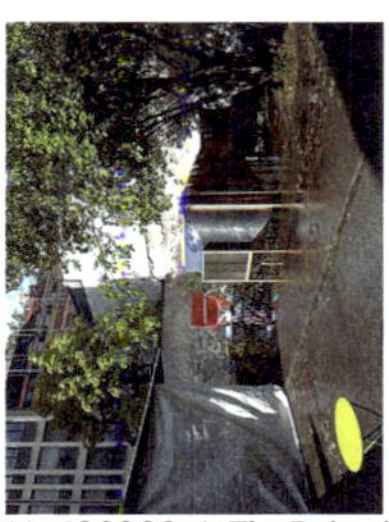
***_130900_InTheRain.tif

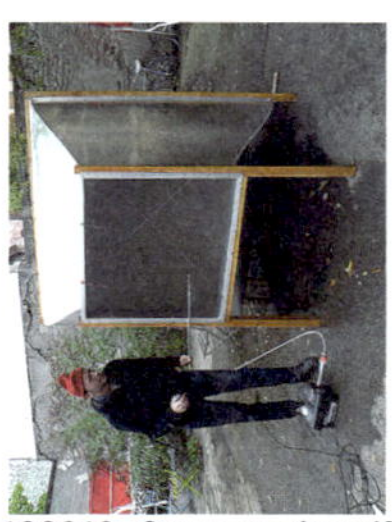
130919_Conversation.tif

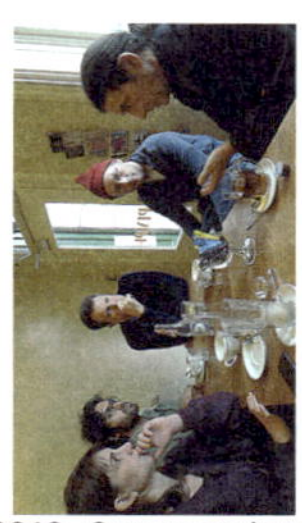
130919_Conversation1.tif

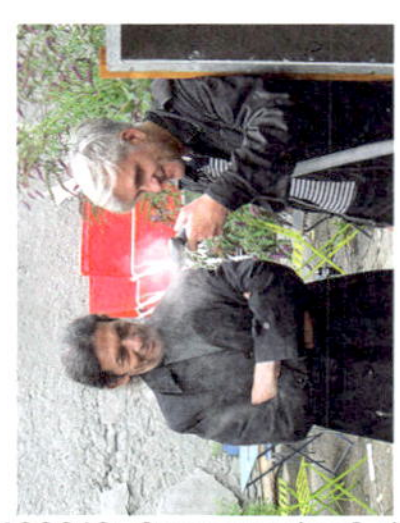
130919_Conversation2.tif

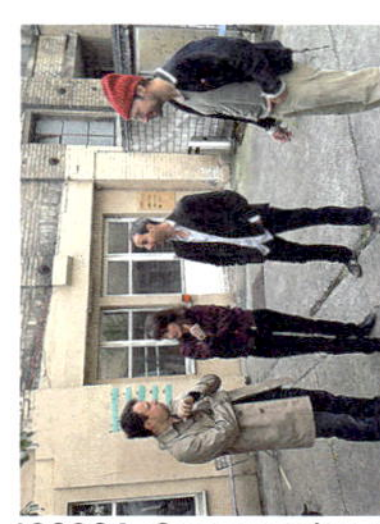
130924_Conversation.tif

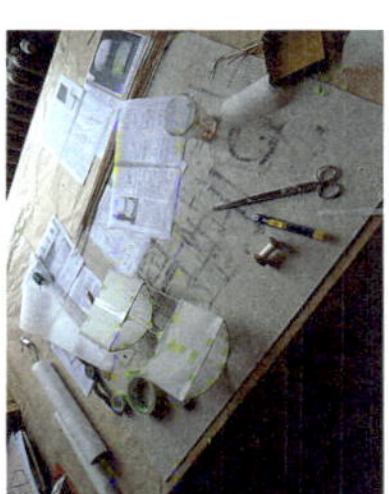
130929_MakingKites.tif

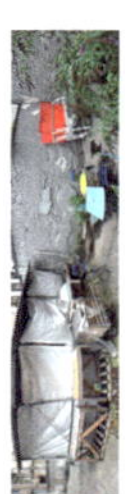
130930_TempleOfWi.tif

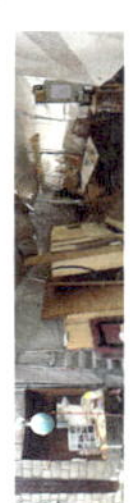
130930_TempleOfWi1.tif

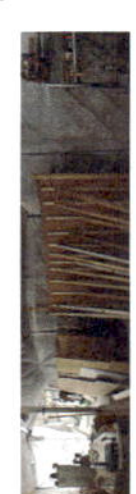
130930_TempleOfWi2.tif

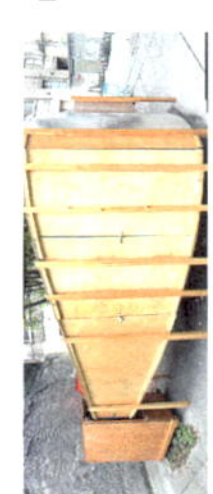
130930_TempleOfWi3.tif

Jul 2013
Curve

Finishing the curve and inlet was the next step in the wind tunnel's construction. The funnel-shaped frame could then suck in the wind from the nozzle and lead it into the second test section in front of the engine. In this way the circular flow (see August 2013) of a Göttingen-type tunnel would become possible. Not only would the wind now circulate, but with the expanded curve, the tunnel was also about to become a habitable space.

130710–130711
Meeting
On the power of writing and the different styles and intentions of technical writing and poetic writing. The team collects poetry on wind.

130710–130711
Conversation
Robert Dennhardt, cultural historian and technology researcher, talks about chaos theory and how it relates to wind. On the second day Ming Liang, Chinese designer and professor from Singapore, joins the wind tunnel tests.

130718
Construction
The curve is ready and all parts are combined to create a new version of the 'Temple of the Winds'.

Aug 2013
Related Shapes

During his summer break, far from Zurich, KK passed by some roof exhaust ventilators made of shiny metal in Jakarta, the capital of Indonesia. The pinnacles seemed vaguely related to the wind tunnel's nozzle (see January 2013). That form had often astounded visitors, reminding them of Baroque church spires.

130829
Multiple
FD starts his series of *Wind Tunnel Drawings*, reprinted in the *Wind Tunnel Bulletin #1* and presented in the group shows 'Round the Corner' and 'less rooms for lies' (see Selected Publications).

130825
Publication
MB participates in the *Conference of the International Cartographic Association* in Dresden with a film.

Sep 2013
Open Research

As an open-air laboratory or an open-air studio, the wind tunnel worked in all types of weather and throughout the seasons. Anyone who was interested could turn on the motor of the tunnel's fan and get acquainted with its wind and its sound.

130909
Construction
The surfaces of the wind tunnel are prepared with spar varnish, protecting it against UV light and water. Now the tunnel is white inside and golden outside, which adds a beautiful glare to the instrument.

130919–130920
Meeting
On modes of working and daily structures and how this influences the work.

130920
Bulletin
Preparation of the first *Wind Tunnel Bulletin* with the designer Viola Zimmermann.

130919
Conversation
Pavel Büchler, artist and professor from Manchester, discusses art and artistic research with the group.

130924
Conversation
Benno Wirz, PhD candidate and assistant at Zurich University, comes to discuss cooperations.

Robert Dennhardt:
The wind tunnel is a physical iteration. However, the bisectrix of the graphical iteration will be drawn only when the wind tunnel has been completed.

Ming Liang:
There is a Chinese proverb: Two monks are looking at a flag in the wind. The first monk says: "The flag is flying." "No", says the second, "it is the wind, which is flying". A third young monk hears it and says: "Excuse me, the mind is flying." Later he became the abbot.

Pavel Büchler:
Is it art? What else could it be!

Benno Wirz:
This is a miracle machine.

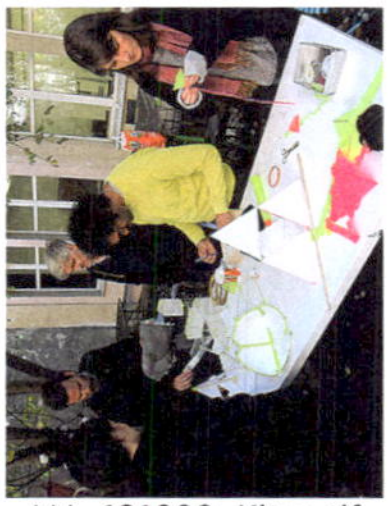
***_131000_Kites.tif

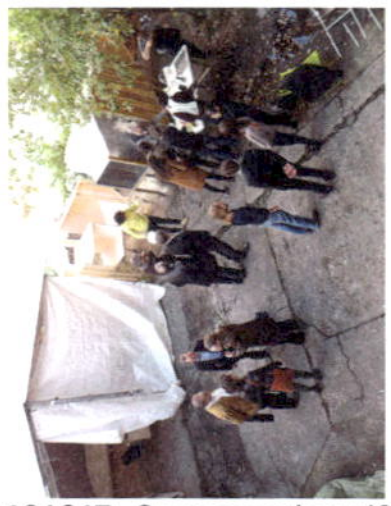
131017_Conversation.tif

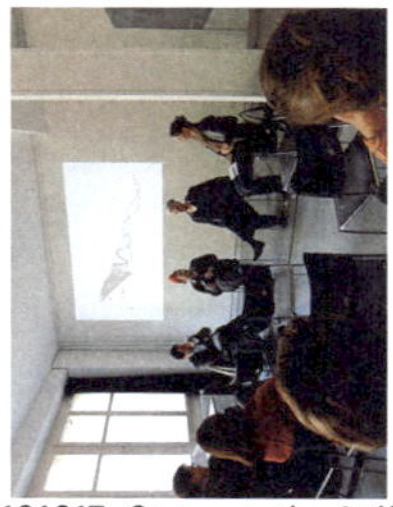
131017_Conversation1.tif

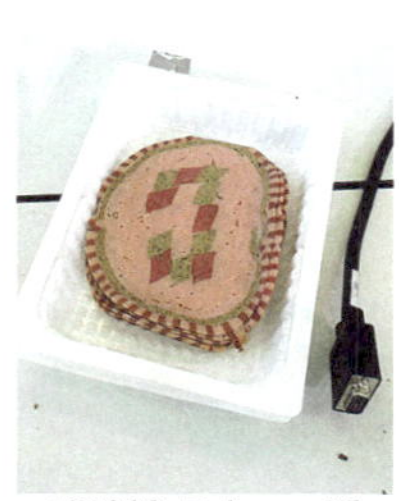
131022_Bologna.tif

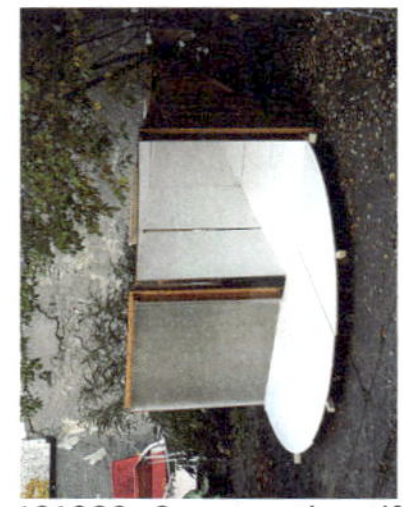
131022_Construction.tif

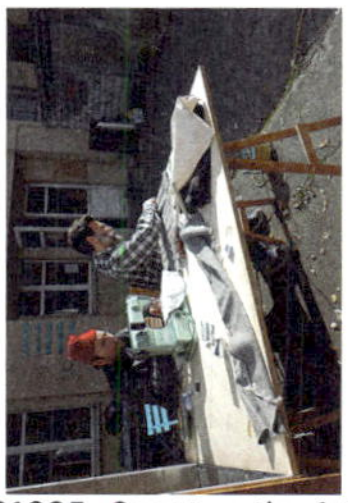
131025_Construction1.tif

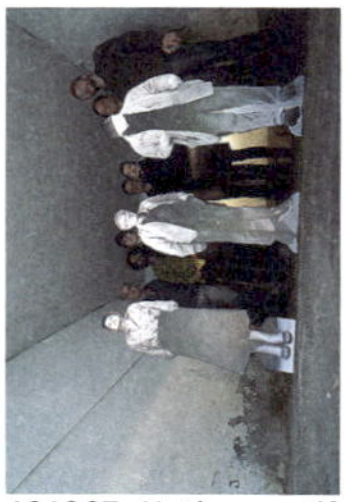
131027_Hothouse.tif

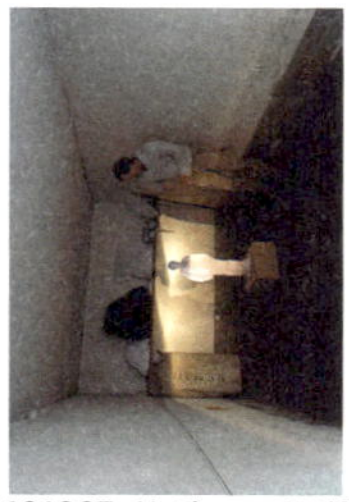
131027_Hothouse1.tif

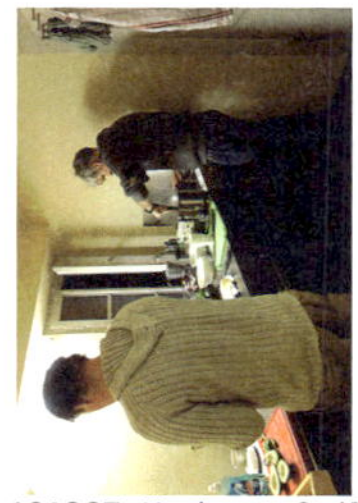
131027_Hothouse2.tif

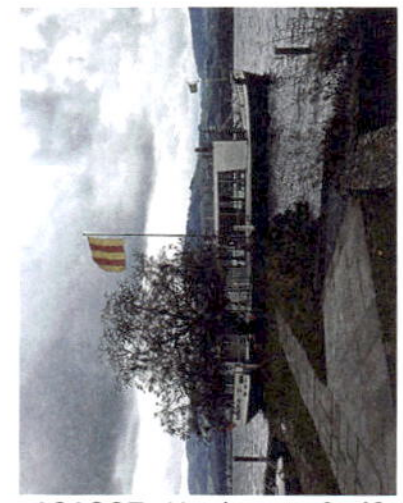
131027_Hothouse3.tif

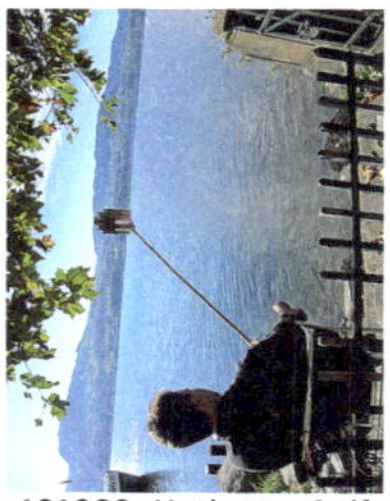
131028_Hothouse4.tif

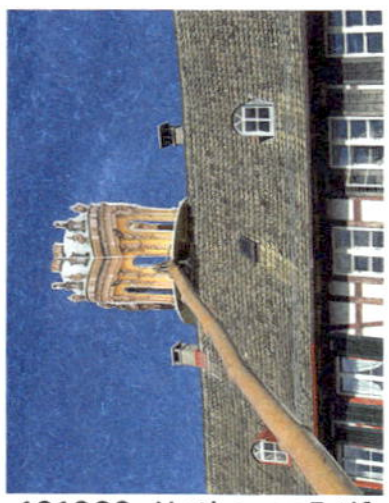
131028_Hothouse5.tif

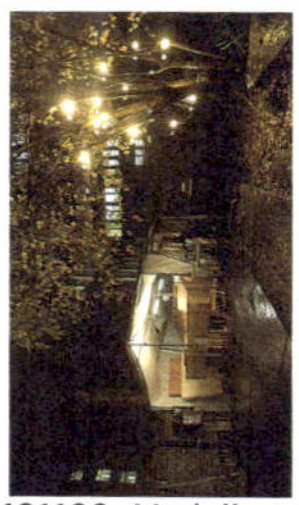
***_131100_Modellapero.tif

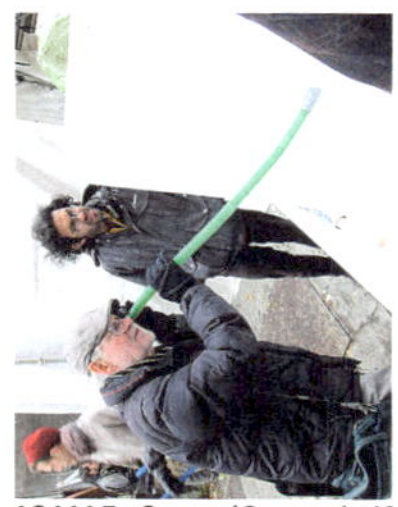
131115_SoundSummi.tif

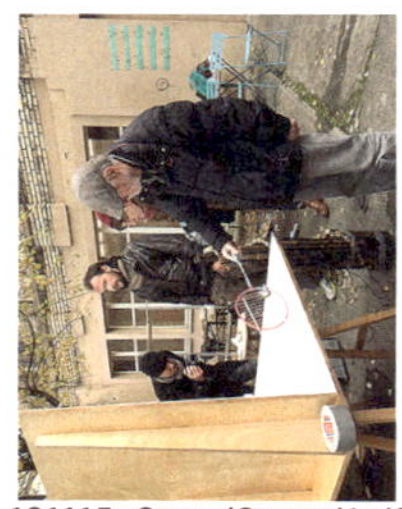
131115_SoundSummi1.tif

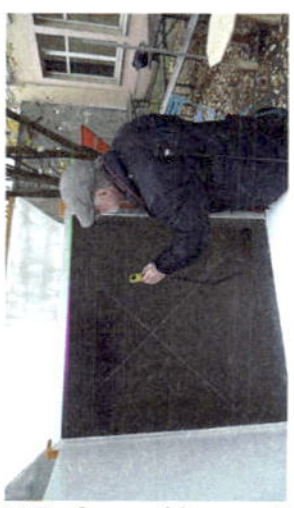
131115_SoundSummi2.tif

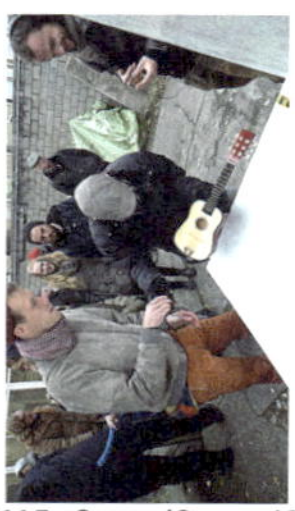
131115_SoundSummi3.tif

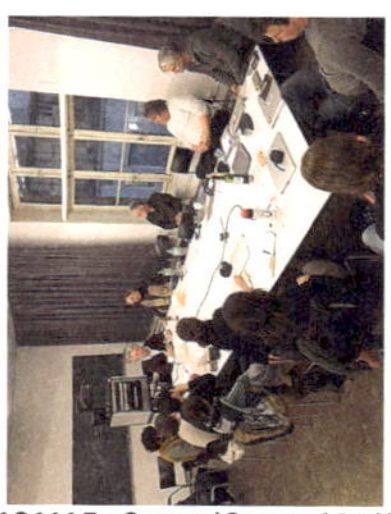
131115_SoundSummi4.tif

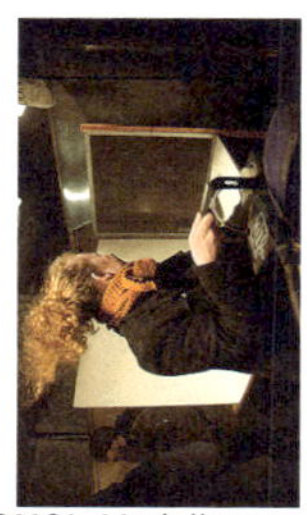
131121_Modellapero.tif

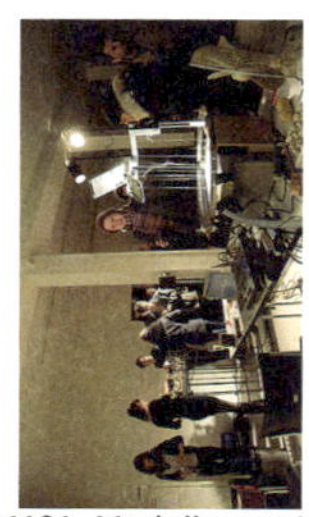
131121_Modellapero1.tif

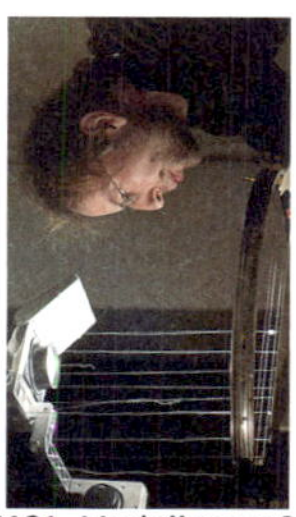
131121_Modellapero2.tif

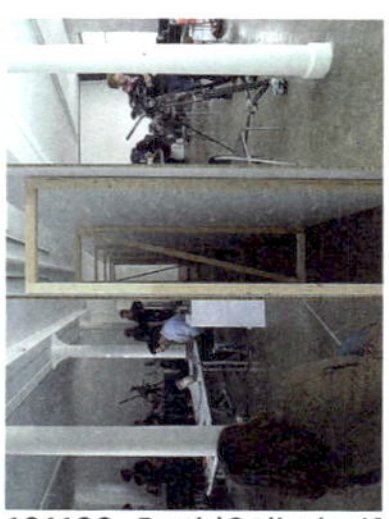
131122_DavidGoliath.tif

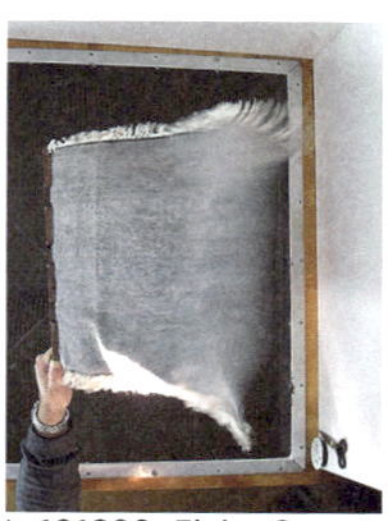
***_131200_FlyingCarpet.tif

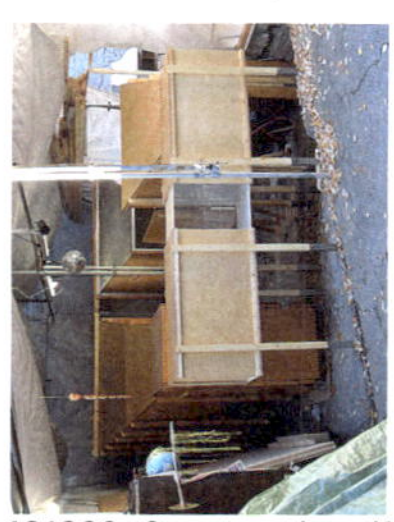
131203_Construction.tif

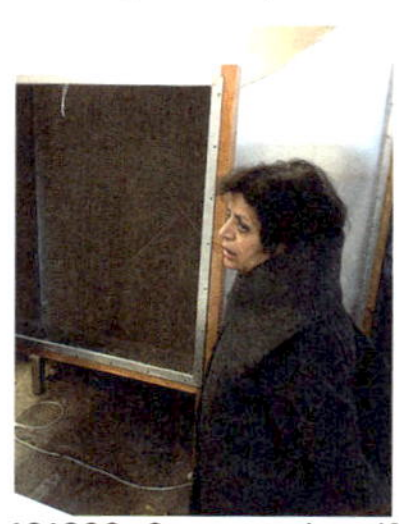
131203_Conversation.tif

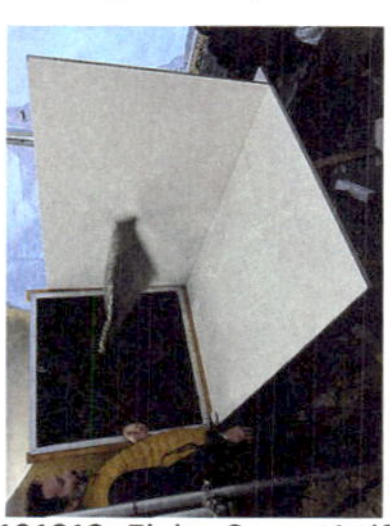
131212_FlyingCarpet1.tif

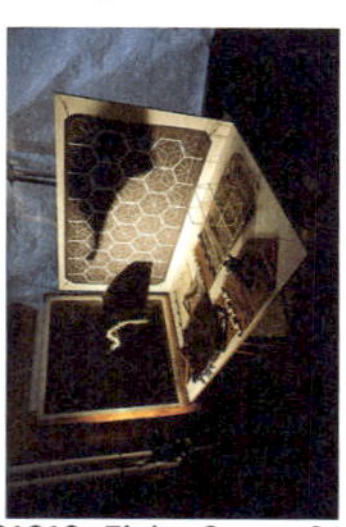
131212_FlyingCarpet2.tif

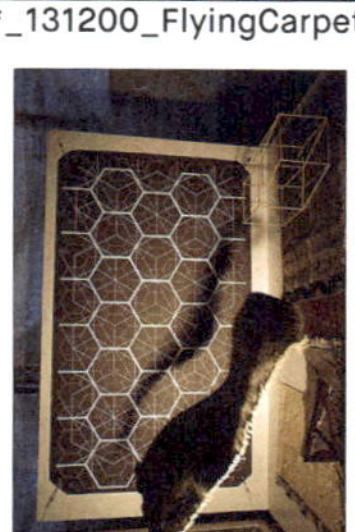
131212_FlyingCarpet3.tif

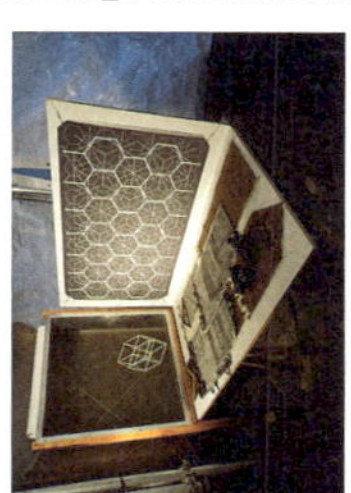
131212_FlyingCarpet4.tif

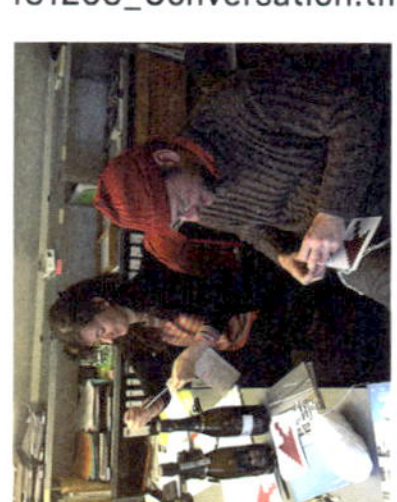
131217_XmasGifts.tif

Oct 2013
Kites

The wind tunnel has a straightening effect on winds. Kites on the other hand use natural non-straight winds, and their diverse shapes bear no relation to the streamlining wind-tunnel world of the 20th century. During the 'Size Matters' group meeting, an assorted collection of kites were constructed and flown in the air at the front of the nozzle. The box kites featuring a tetrahedron structure prevailed. This kite format, often both made up of tetrahedrons and in itself a tetrahedron, is regarded as one of the most stable structures and owes its design to the inventor of the telephone, Alexander Graham Bell.

131015
Artist in Residence
Luis Berríos-Negrón starts his residency with the 'Size Matters' project. He uses the wind tunnel for his research on flying carpets (see December 2013). The films are later shown e.g. at the Bahia Biennial in Brazil.

131017
Meeting
The team builds kites.

131017
Conversation
The scientific fellows of Collegium Helveticum come to visit the wind tunnel with their director and PhD students. RW presents his new book *Modelle in Kunst und Wissenschaft* (Models in Art and Science; see Selected Publications) and a podium discussion is chaired by Ladina Bezzola.

131022
Multiple
In preparation for the *Modellapéro*, a mortadella bologna with a wind tunnel pattern is designed and tested.

131026–131102
Hothouse
The whole team spends a week in Uerikon, next to Lake Zurich. The week is dedicated to the topic of models. Activities include: lectures, performances, verbal and non-verbal exercises. All these help to form a new sensibility for group work (see September 2015). A documentry film is made that records the activities of the research week (see Selected Publications).

Nov 2013
Modellapéro

The *Modellapéro* was initiated as a format for live publication and performance. The group presented their work directly in the environment where they had been working. The public had the opportunity to both experience these interdisciplinary activities and to pose questions.
At the second *Modellapéro* (the first was in November of 2012) the intensity of the crossover and integration between the cultural historians, the artists and the public, was explicit: all guests were invited to bring their own model and test it in the wind tunnel. The film *The Endless (Wind) Tunnel* made from found-footage by FD and SW deals with representations of flow – developing a generic impression of wind tunnel tests.
It was screened in a little cinema in the main building while, outside, the real tunnel was testing new objects. KK played live electronics and the wind harp.

131115
Conversation
'Sound Summit' including Peter Kiefer, sound artist and professor in Mainz, with his students; Anthony Moore, sound artist and professor at the Academy of Media Arts Cologne, also with students; Max Eastly, sound artist from London; and Andres Bosshard, sound artist and professor at ZHdK. The summit is dedicated to experiments with the sound of the wind. The composer Alvin Curran, who is in Zurich for a concert, joins the talks.

131121
Construction/Modellapéro
The closed Göttingen-type loop is inaugurated and presented to the public.

131121
Bulletin
The Wind Tunnel Bulletin #1 is launched.

131121
Multiple
A flat, arrow-shaped Japanese pen is included with the Bulletin #1. At the Modellapéro the team is wearing the 3D-printed pink pins of the wind tunnel.

131122
Publication
An international symposium is held under the title *David and Goliath: Models between Art and Architecture* with Luis Berríos-Negrón, artist, Berlin; Ellen Blumenstein, KW Institute for Contemporary Art, Berlin; Vera Bühlmann, Laboratory of Applied Virtuality, Swiss Federal Institute of Technology Zurich; Teresa Fankhänel, University of Zurich; and Anja Sattelmacher, Humboldt University, Berlin. Organized by FD, RW and Stephan Trüby, architect and head of the Master of Advanced Studies in Spatial Design at ZHdK.

Dec 2013
Flying Carpets

Artist in residence Luis Berríos-Negrón brought prototypes of his flying carpets and tested them in the wind tunnel. The flying carpets represent the dream of returning to lost homelands, among other things. Hence the carpets introduced the dimension of melancholy to the testing.
Berríos-Negrón also worked with several screens with tesseract ornament, which were subsequently used as a background for most tests. Berríos-Negrón recorded a series of slow-motion films featuring the shadow of the carpets. The tesseract and the shadows point to ways of assisting the understanding of a fourth dimension (see January 2014). Perhaps all the wind tunnel tests are approaches to the fourth dimension?

131203
Construction
Luis Berríos-Negrón builds the test section display table.

131203
Conversation
Vishakha Desai, professor from New York and Asia scholar as well as adviser on Global Affairs, visits the tunnel and discusses wind and globalization.

131205
Meeting
Wrap up of the year, look ahead to 2014 and a review of the Hothouse in Uerikon.

131206–131207
Publication
HA and SW present their PhD work for the first time at a conference, held at LaborARTorium, Maximilian University in Munich.

131215
Artist in Residence
Luis Berríos-Negrón concludes his residency.

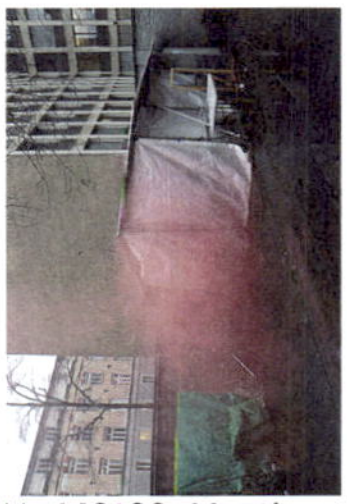
***_140100_Meeting.tif

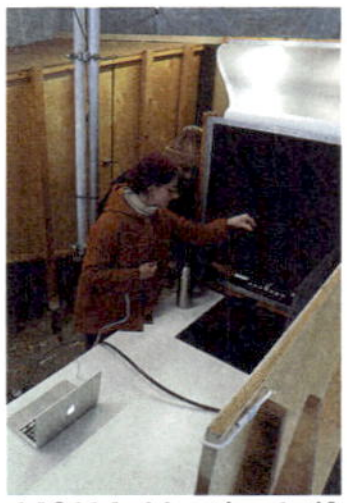
140114_Meeting1.tif

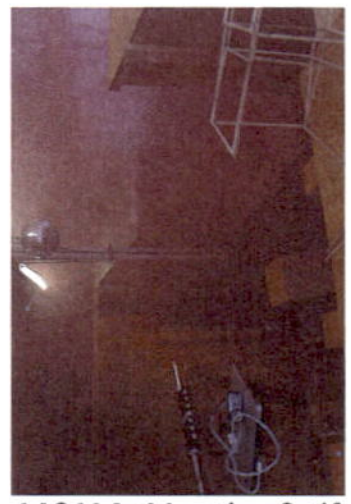
140114_Meeting2.tif

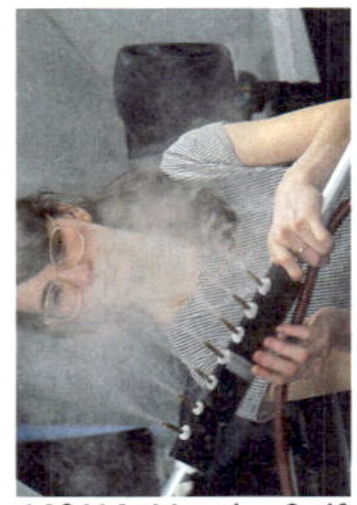
140114_Meeting3.tif

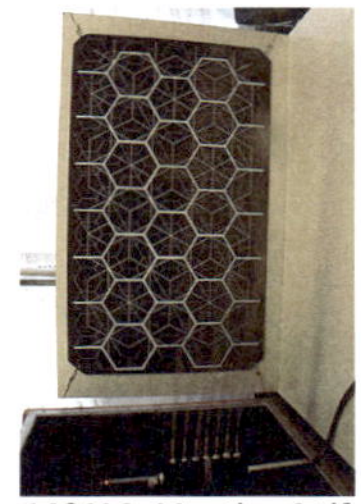
140114_Meeting4.tif

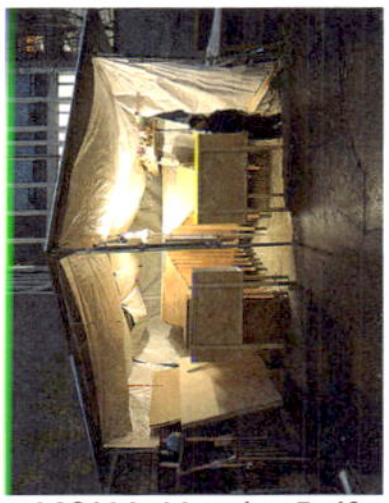
140114_Meeting5.tif

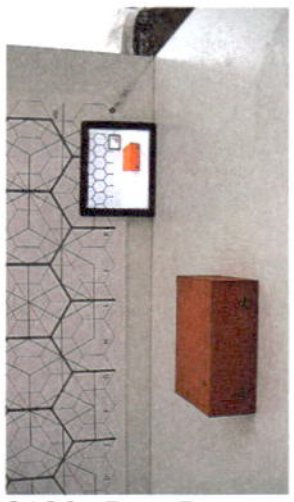
140129_PrepResear.tif

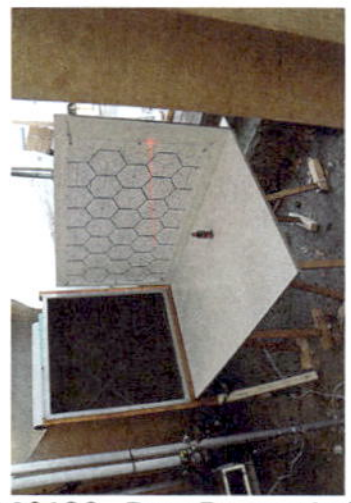
140129_PrepResear1.tif

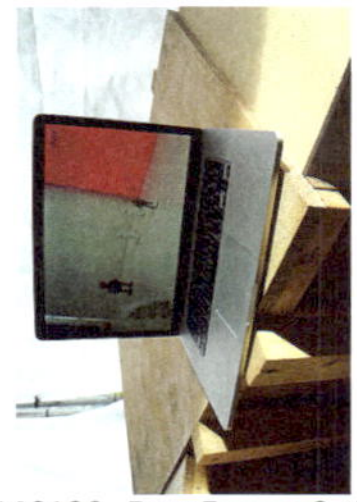
140129_PrepResear2.tif

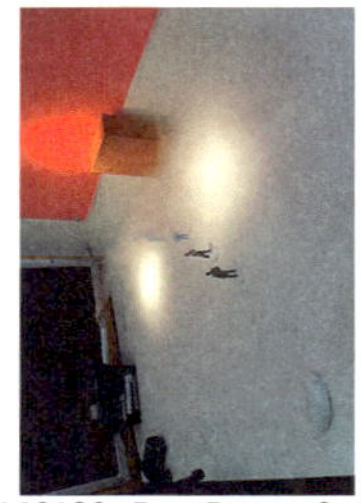
140129_PrepResear3.tif

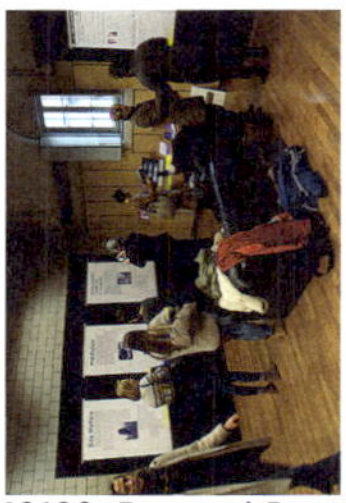
140130_ResearchDay.tif

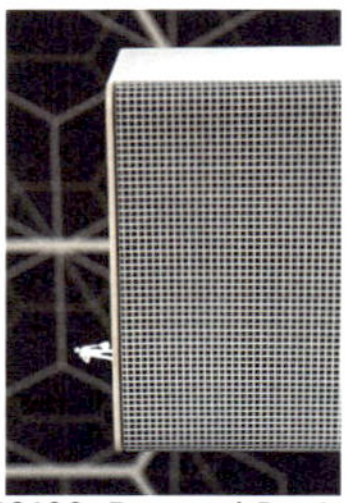
140130_ResearchDay1.tif

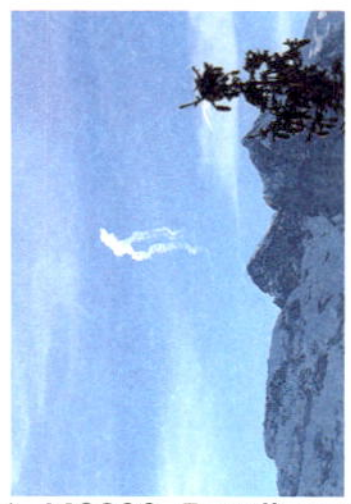
***_140200_Engelberg.tif

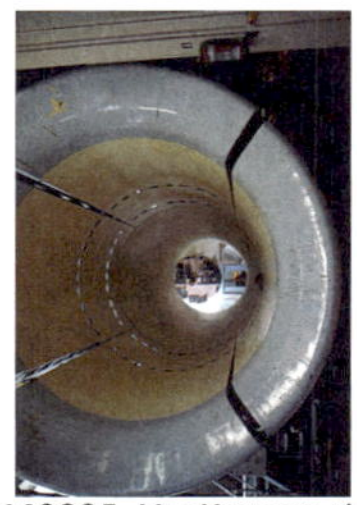
140205_VonKarman.tif

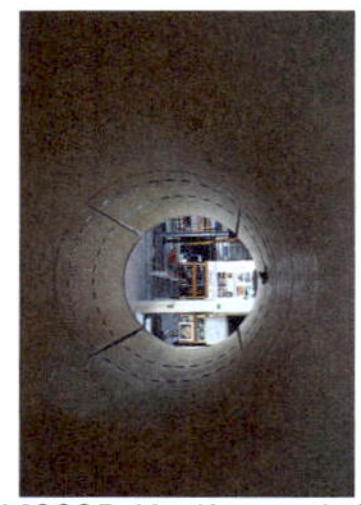
140205_VonKarman1.tif

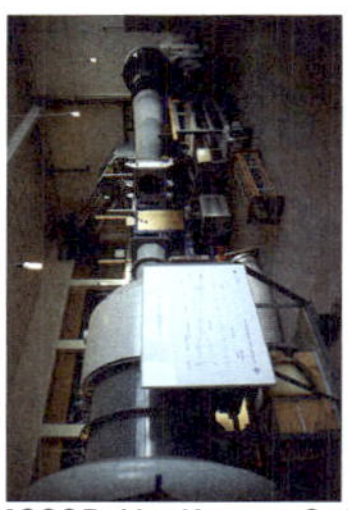
140205_VonKarman2.tif

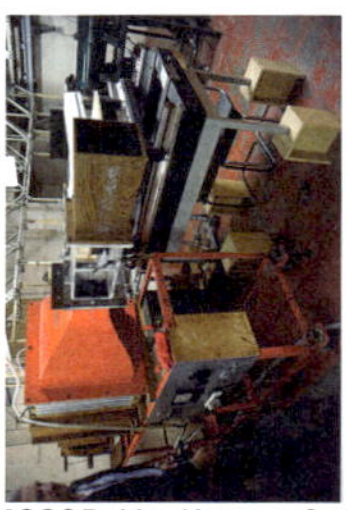
140205_VonKarman3.tif

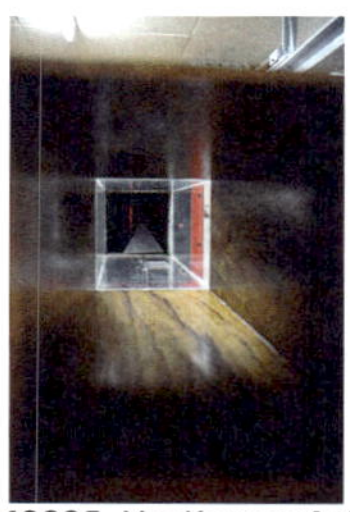
140205_VonKarman4.tif

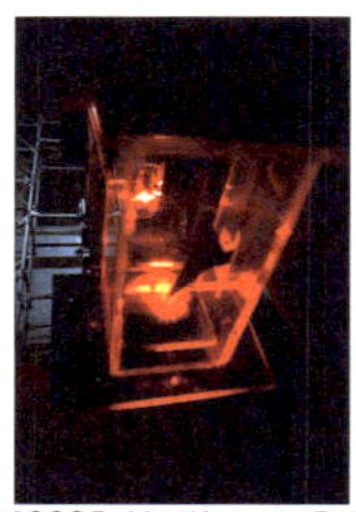
140205_VonKarman5.tif

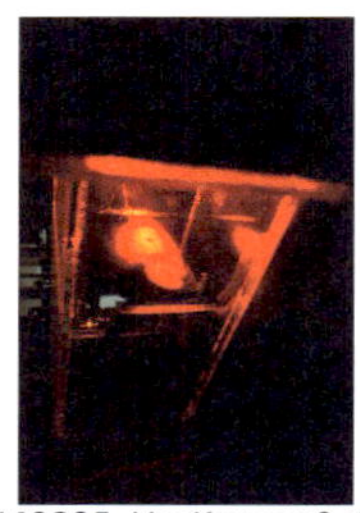
140205_VonKarman6.tif

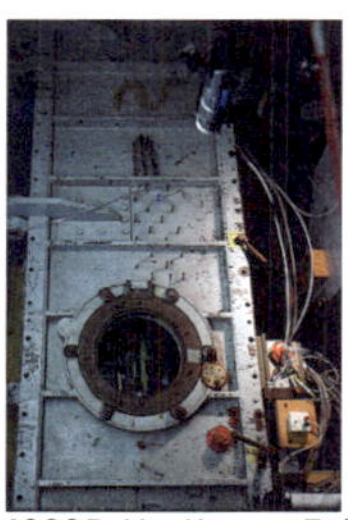
140205_VonKarman7.tif

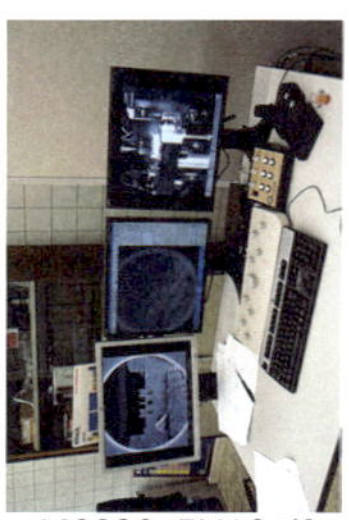
140206_EMAC.tif

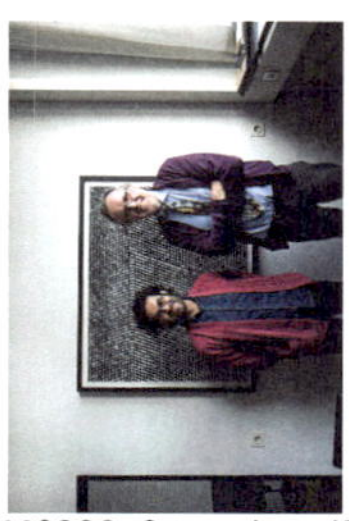
140206_Supervisor.tif

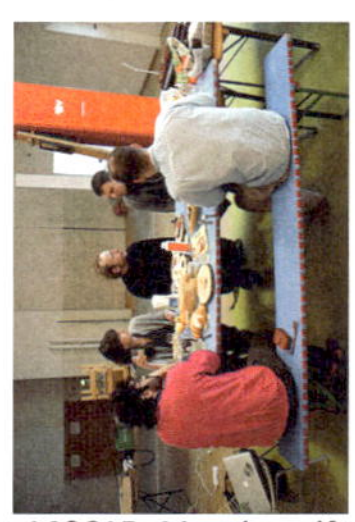
140215_Meeting.tif

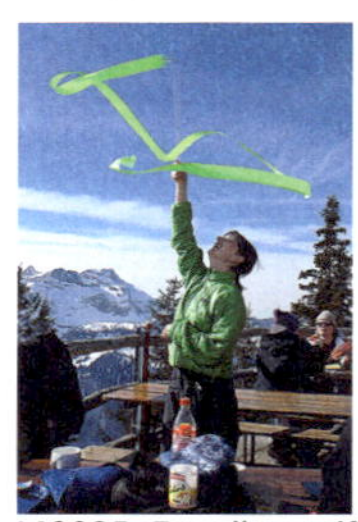
140225_Engelberg.tif

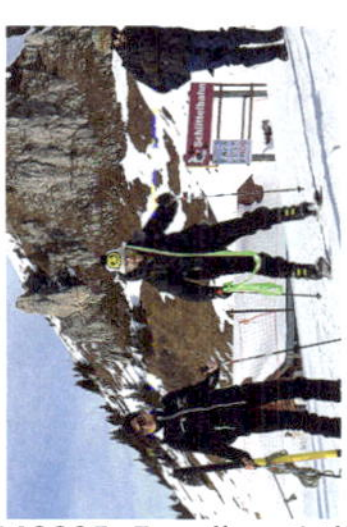
140225_Engelberg1.tif

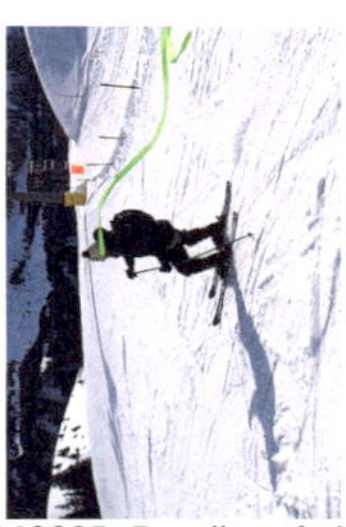
140225_Engelberg2.tif

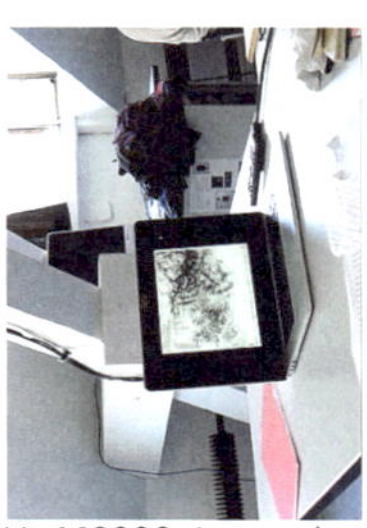
***_140300_Leonardo.tif

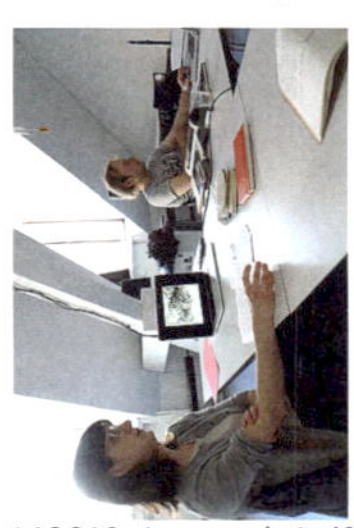
140313_Leonardo1.tif

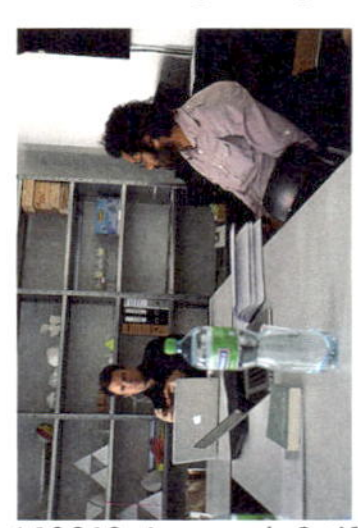
140313_Leonardo2.tif

Jan 2014
Smoke

Smoke is a key partner to the wind, as it enables its display without the use of auxiliary material objects. Smoke is used to describe the wind's flow and its intensity. It can detect invisible openings and then disperse into the air. In the Hafnerstrasse yard, the fumes did not draw that much attention before they disappeared. This would change when the wind tunnel moved to the Toni Areal. There, despite the inoffensiveness of the smoke used in the wind tunnel experiments, it often aroused instinctive nervousness because of its association with fire (see October 2014). The first extensive smoke-tests included colour as a component.

140113–140115
Conversation
Mariel Villeré, curator, art historian and assistant from the partner group at MIT Boston, visits the tunnel to discuss the continuation of teaching exchanges with the ACT programme.

140114
Meeting
Team exchanges on working techniques and smoke.

140130
Publication
At ZHdK's Research Day the wind tunnel is present through a Skype connection, inviting people in the main building to walk the 15 minutes to the working tunnel.

Feb 2014
Traces

On the second skiing excursion to Engelberg, the group tested large 'tell-tale' ribbons and a very large tail appeared in the sky. Can thought manipulate reality? FD: *"Sometimes there are these beloved but strange moments: when reality seems to prove what you have just said. And you wonder, do you see reality differently because of what you are saying? Or does reality really realize what you've just said?"*

140205–140206
HA's PhD committee is selected and the members of the jury visit the Von Karman Institute for Fluid Dynamics outside Brussels as well as the EMAC research group at the University of Antwerp.

140215
Meeting
Team meets in Bern. At the meeting the team collects quotes and images on scaling and also prepares for the second *Wind Tunnel Bulletin*.

140225
Excursion
Skiing in Engelberg, ribbons tested in the wind.

Mar 2014
Writing

Diverse forms of writing criss-crossed in and out of the project. The visual was caught in the flow of words and the words created new images. Scientific articles, spontaneous reflections, fictional texts and writing in preparation for the team's workshops all facilitated different forms of reflection. Furthermore, the *Wind Tunnel Bulletin* cultivated an assemblage of styles and functioned as a receptacle for diverse content.

140313
Meeting
The art historical technique of the verbal description of artworks is studied. Everybody writes about a wind drawing by Leonardo da Vinci.

140327
Teaching
FD lectures on models and scale via Skype to a class at Johns Hopkins University in Baltimore and tests remote sensing with MB.

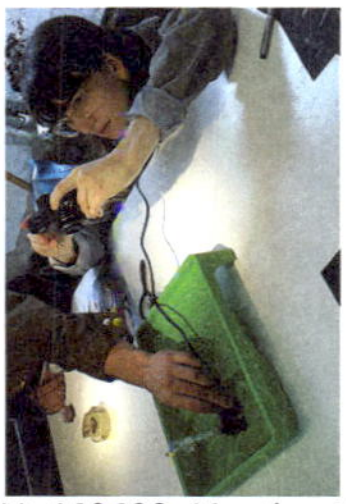
***_140400_Meeting.tif

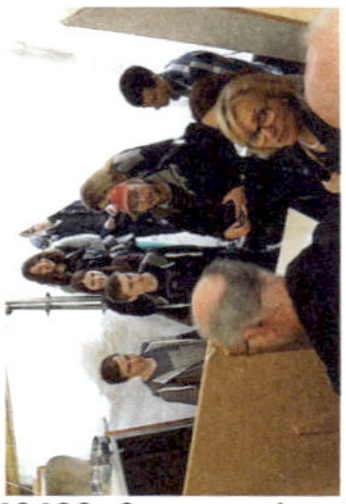
140408_Conversation.tif

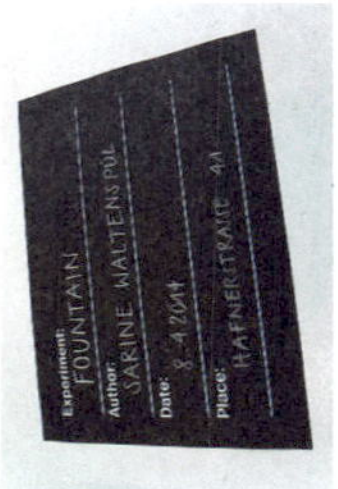

140408_Meeting1.tif

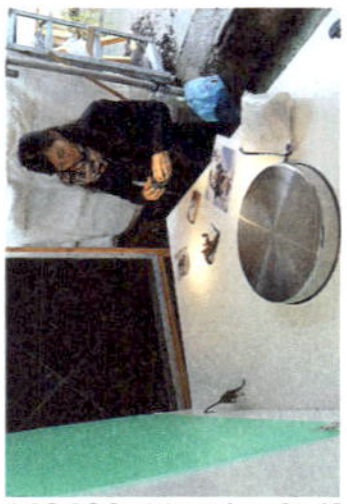
140408_Meeting2.tif

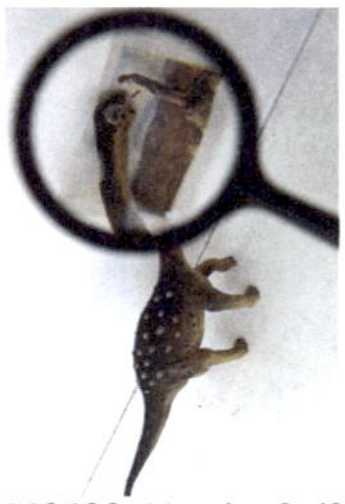
140408_Meeting3.tif

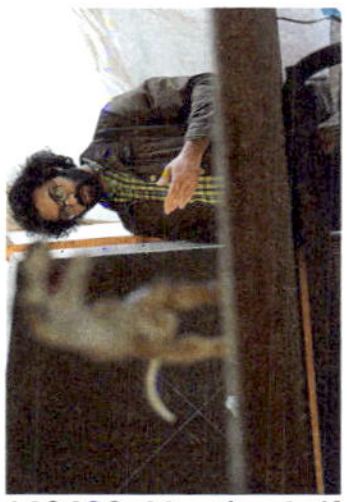
140408_Meeting4.tif

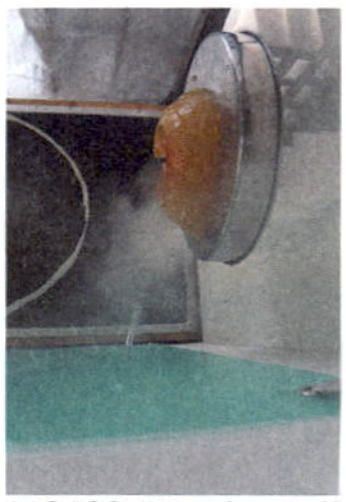
140408_Meeting5.tif

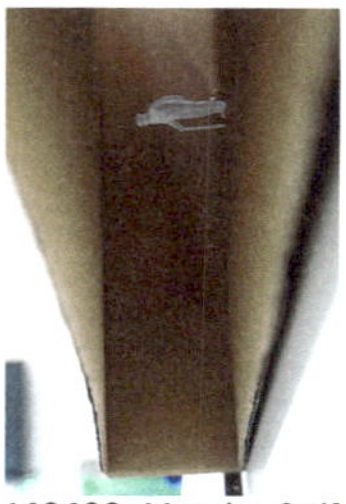
140408_Meeting6.tif

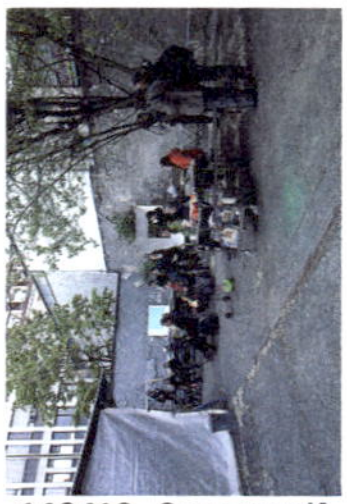
140416_Concert.tif

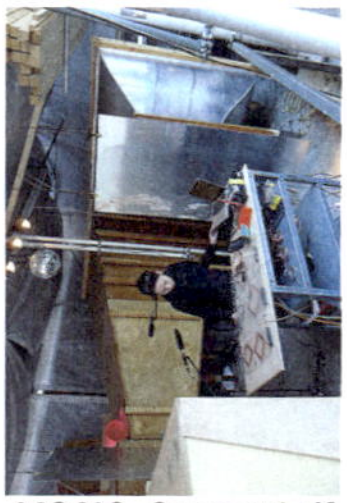
140416_Concert1.tif

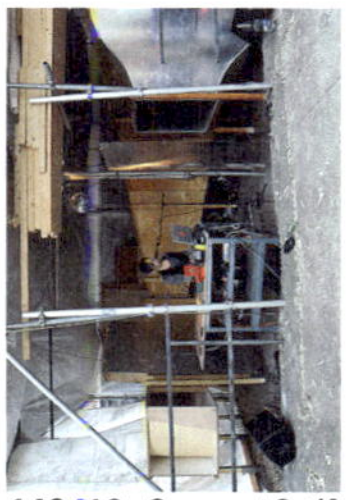
140416_Concert2.tif

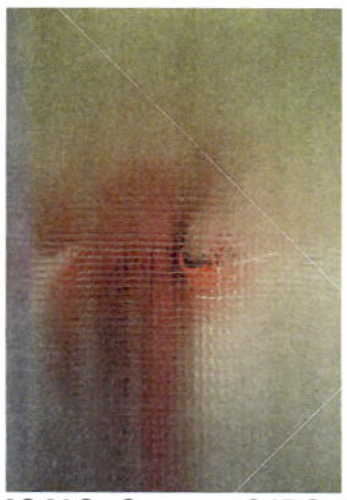
140416_Concert3JPG.tif

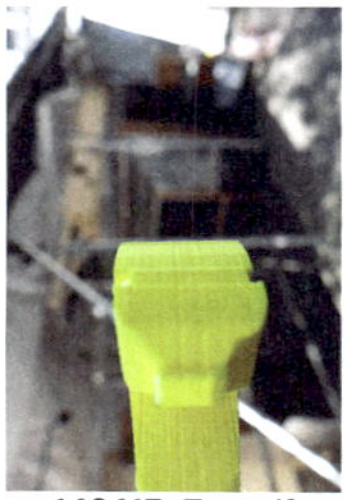
140417_Test.tif

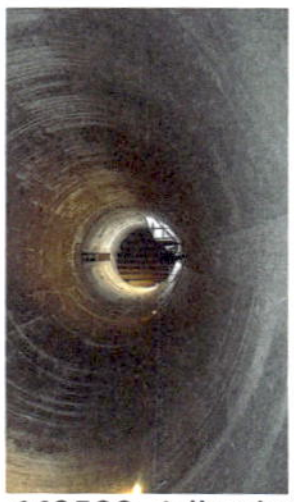
***_140500_Adlershof.tif

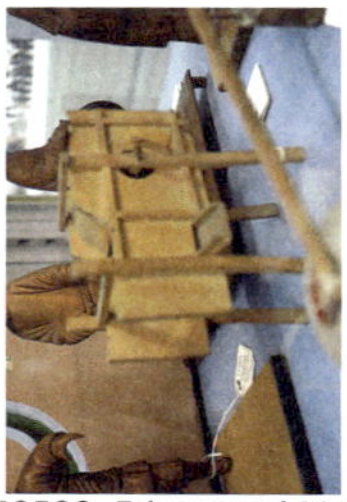
140508_EthnographM.tif

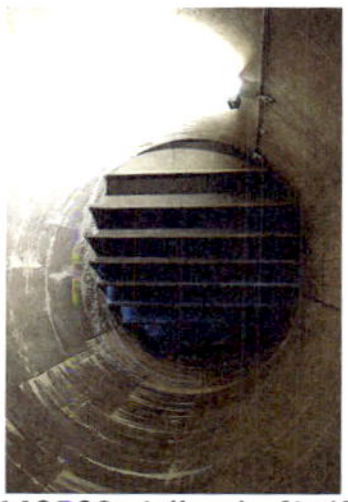
140509_Adlershof1.tif

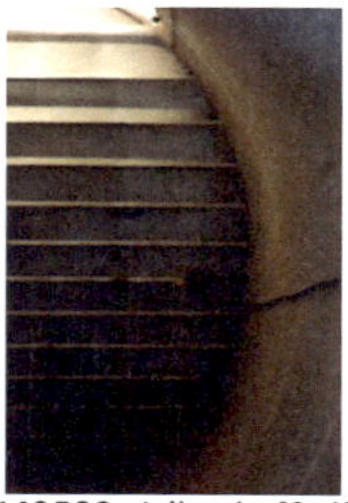
140509_Adlershof2.tif

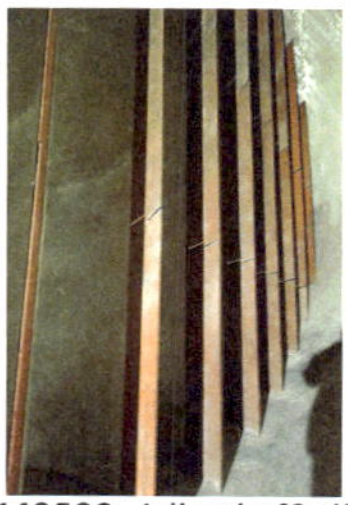
140509_Adlershof3.tif

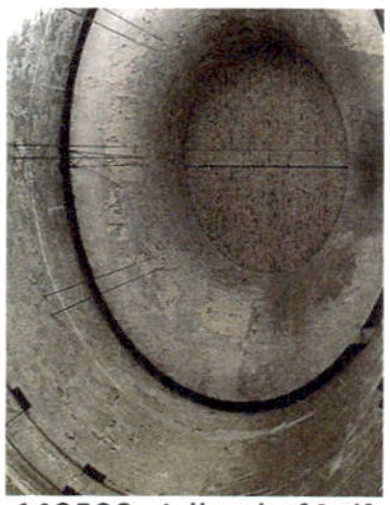
140509_Adlershof4.tif

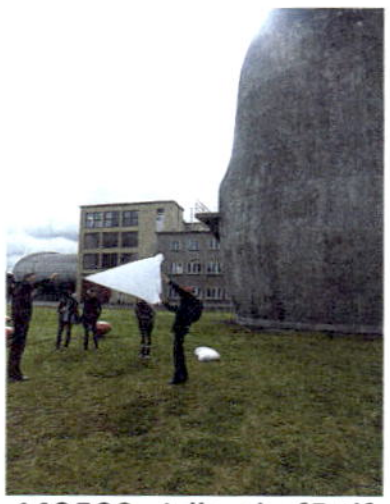
140509_Adlershof5.tif

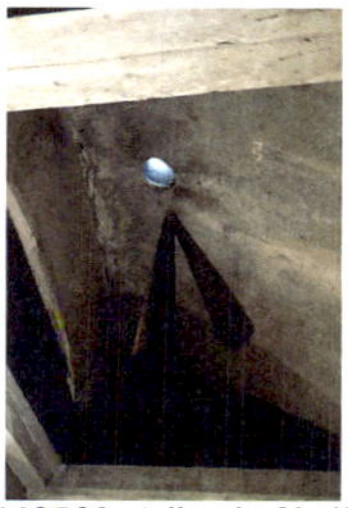
140509_Adlershof6.tif

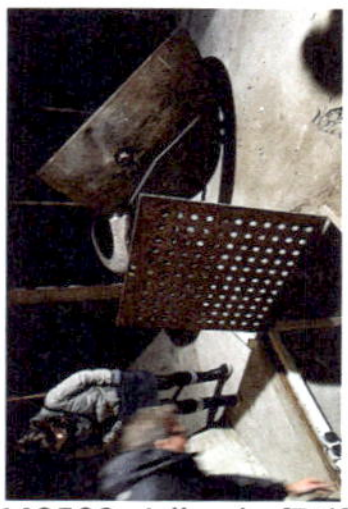
140509_Adlershof7.tif

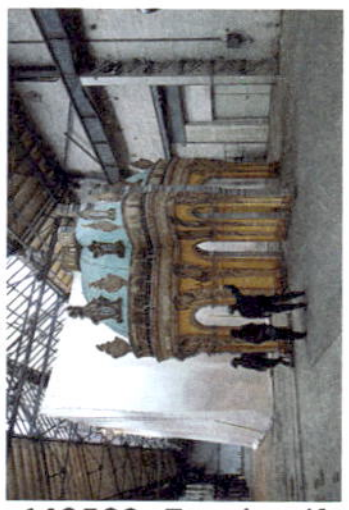
140509_Zugabe.tif

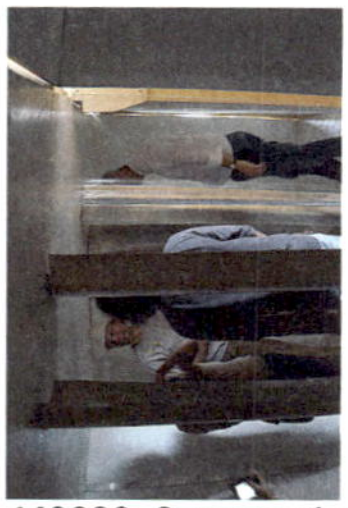
***_140600_Conversation.tif

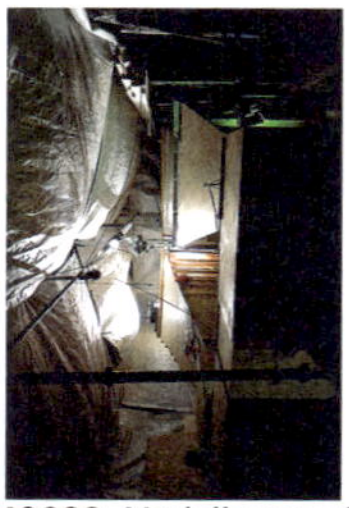
140626_Modellapero.tif

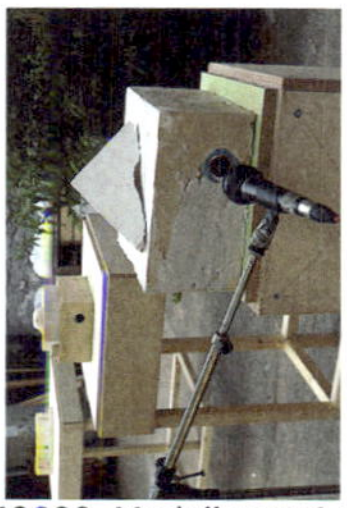
140626_Modellapero1.tif

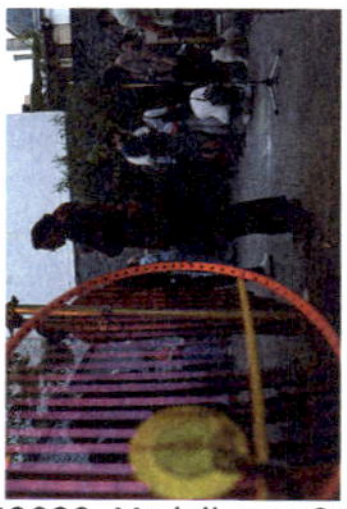
140626_Modellapero2.tif

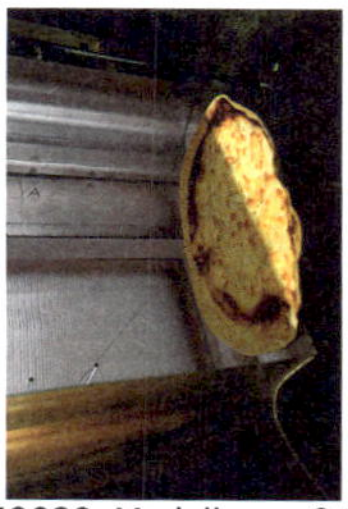
140626_Modellapero3.tif

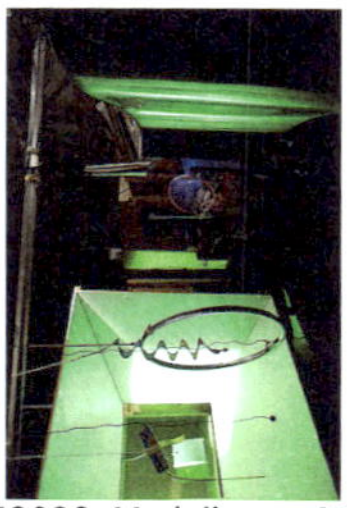
140626_Modellapero4.tif

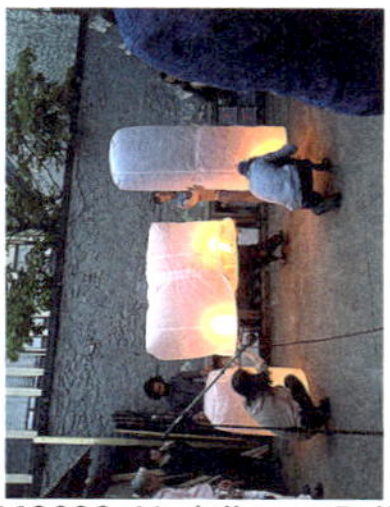
140626_Modellapero5.tif

Apr 2014
Scaling

The group experimented with cameras and frame rates to make the process of using cameras for scaling visible and to test the scale of the models in the wind in front of the nozzle. When it comes to visual deception, water does not collaborate smoothly with the lenses, as you cannot depict the sea by making waves in a barrel. SW: *"There are several ways to create the illusion of greater scale in filming water. The most used is to film with a higher frame rate so the water appears 'slower' when projected. We get the impression of greater mass and thereby the impression of greater scale. So our fountain might have looked bigger filmed in slow motion, but on the other hand, the waves on the water surface were out of scale. For this issue there are other solutions in film making, such as fans and chemicals to influence the surface tension."*

140402–140625
Conversation
KK invites students to spend time working and discussing the wind tunnel with him on Wednesdays, between 10:30 and 12:30.

140408
Meeting
The group tests diverse models and documents them.

140408
Conversation
Gerd Folkers, director of Collegium Helveticum, visits the tunnel with a group of students from Germany.

140416
Conversation
Alexis Emilianoff, a Canadian sound artist, visits and gives a concert in the test section of the wind tunnel.

May 2014
Adlershof

The visit to the Adlershof's main wind tunnel was an encounter with a ghost-like space that stirred echoes of earlier tests and generated prolonged sounds. The physical presence of this empty stage, where science and world politics had once been closely intertwined, is impressive. Then, as now, the military has always had a vested interest in aerodynamics, and the results of tests here also ended up on the battlefield. There are larger tunnels than the Adlershof; however, this one is accessible to the public (on request). Its walled-up nozzle doesn't let any wind pass through it. It is a relic, a scientific dinosaur.

140503
Writer in Residence
Sofie van Loo, curator from Belgium, starts her residency at the wind tunnel.

140508–140509
Excursion
The team travels to Berlin and visits the Aerodynamic Park Adlershof again, the Teufelsberg, the Ethnographic Museum for models and *Zugabe*, FD's work in progress in Potsdam, among others.

Sofie van Loo:
My idea of a wind tunnel is a bi-modern and abstract-realist wind tunnel, which means an instrument for a gentle jumping between 'imagination' – also as a thinking-through of (non-) knowledge – and 'thinking' (of knowledge) without falling into the gap of either (re-)presentation or transformation.

Jun 2014
Evaluation

As an example of a transdisciplinary evaluation Olivier Chazot of the Von Karman Institute for Fluid Dynamics, located outside of Brussels, came to discuss the artistic wind tunnel in Zurich and tested it with a classic tufted wand similar to tell-tales. The artistic image of the wind tunnel was re-encountered by the scientific gaze. HA: *"Maybe it is a bit like a reinvention of the wind tunnel. Like drawing an object that you only know from memory, and that is also what makes it interesting. It is a reflection on another understanding of the sciences."*

140603
Conversation
Olivier Chazot, director of the Aeronautics and Aerospace Department at the Von Karman Institute, evaluates the wind tunnel.

140626
Modellapéro
The last *Modellapéro* before the wind tunnel is moved to the new premises is also a welcome to the mosque and community centre which will be the new residents in the Hafnerstrasse.

140426
Bulletin
The *Wind Tunnel Bulletin #2* is launched with the title *The Scale Effect*.

140426
Multiple
A black-and-white poster of an oversized fruit fly is produced and distributed with the *Wind Tunnel Bulletin #2*.

140628
Writer in Residence
The Belgian curator Sofie van Loo ends her residency.

140630
Construction
The move to the new site at the Toni Areal starts.

140630
Team
RW leaves the team to take up a position as postdoc at the Institute for Art History in Florence.

Olivier Chazot:
We are all different. So why don't we have different wind tunnels?

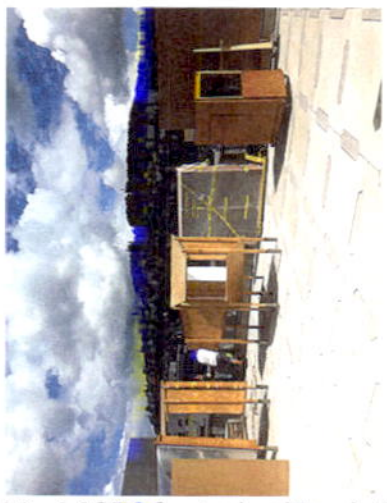
***_140700_ArrivalToni.tif

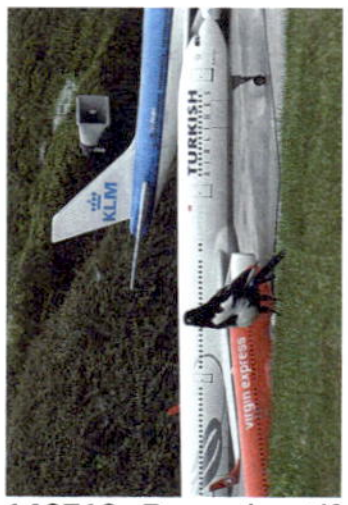
140710_Excursion.tif

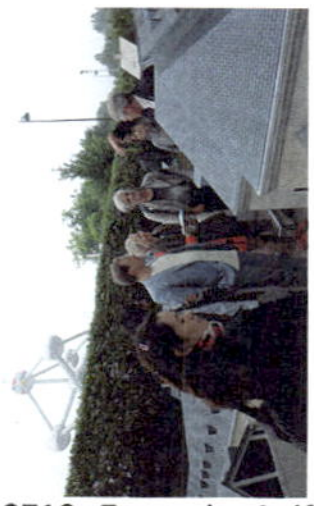
140710_Excursion1.tif

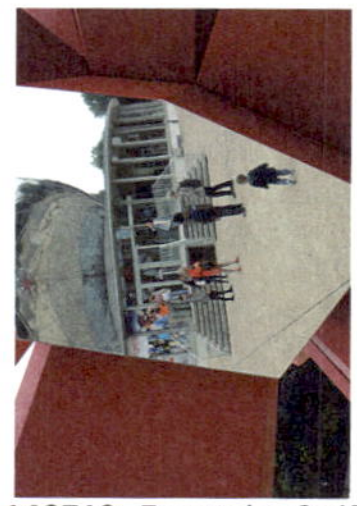
140710_Excursion2.tif

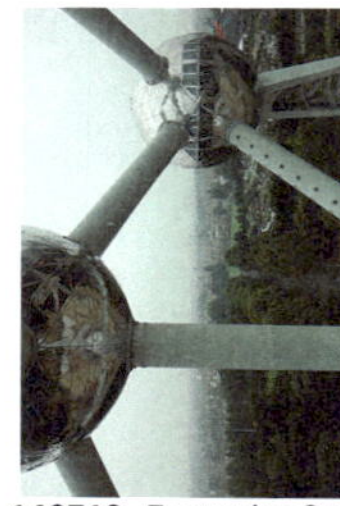
140710_Excursion3.tif

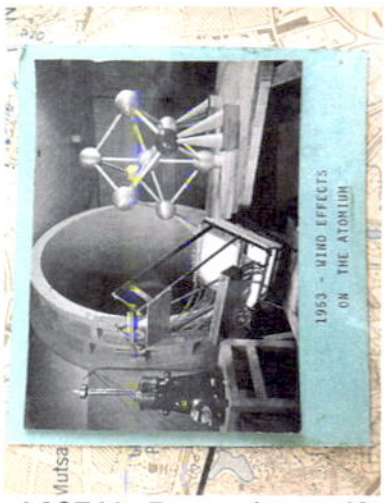
140711_Excursion4.tif

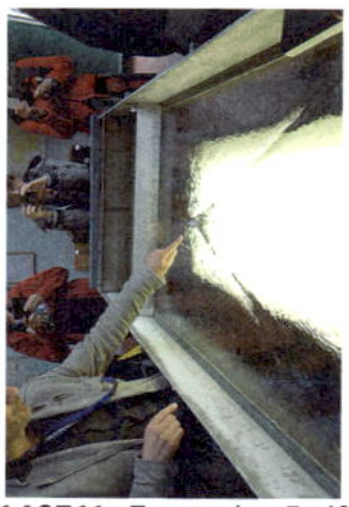
140711_Excursion5.tif

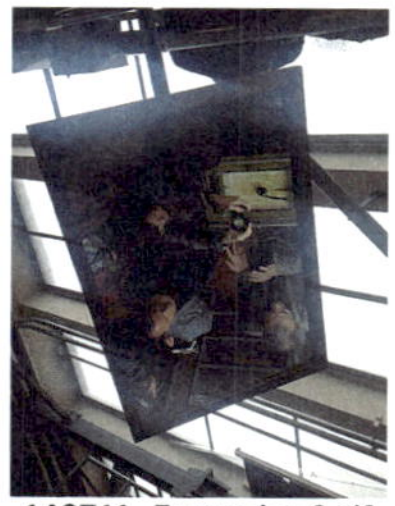
140711_Excursion6.tif

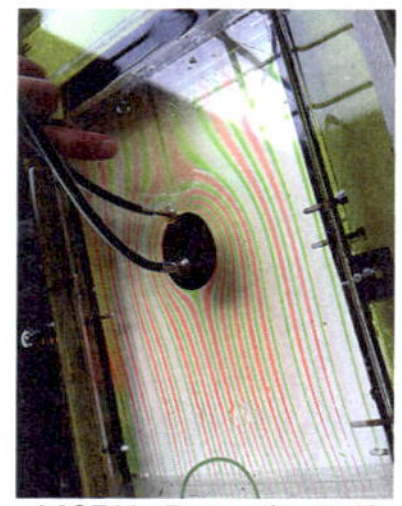
140711_Excursion7.tif

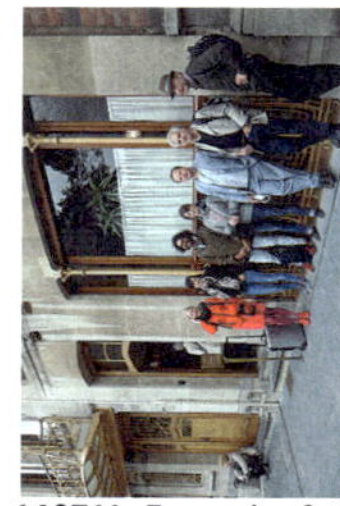
140711_Excursion8.tif

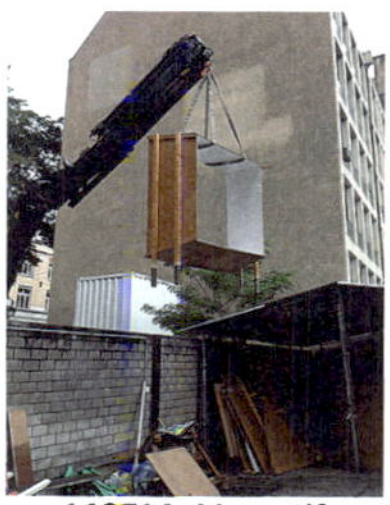
140714_Move.tif

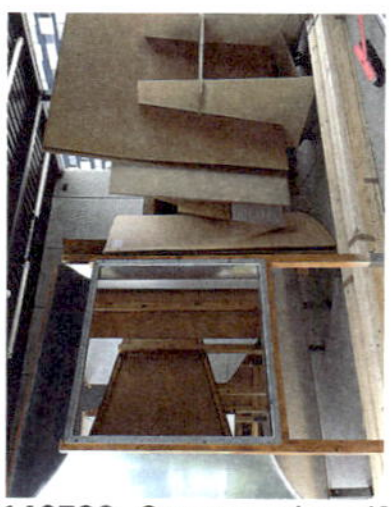
140722_Construction.tif

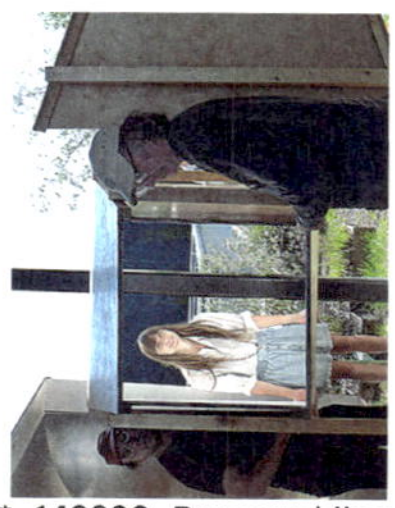
***_140800_Reassembling.tif

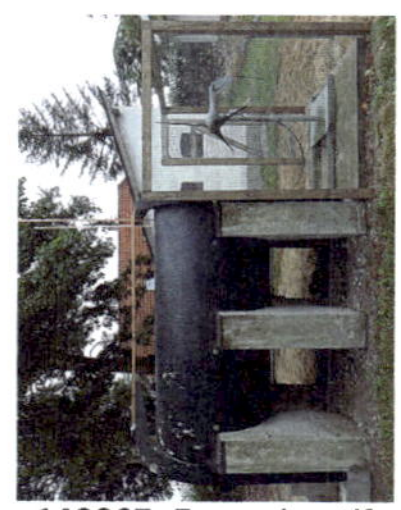
140807_Excursion.tif

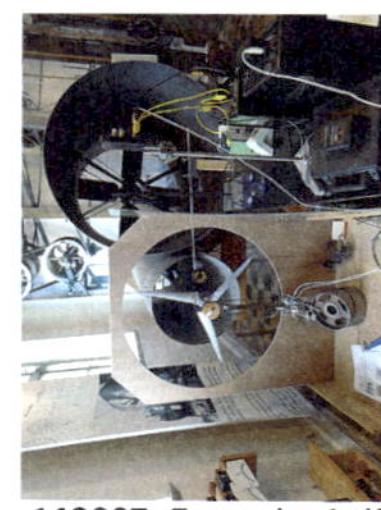
140807_Excursion1.tif

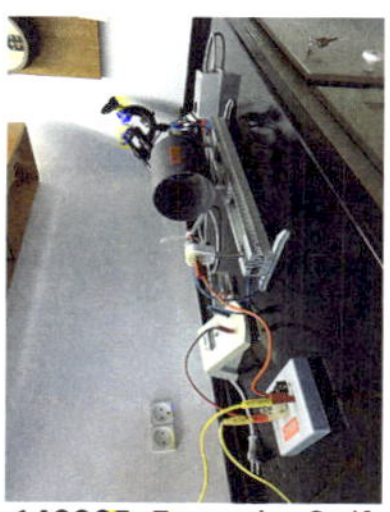
140807_Excursion2.tif

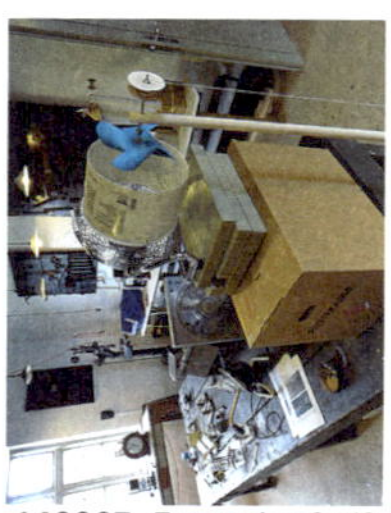
140807_Excursion3.tif

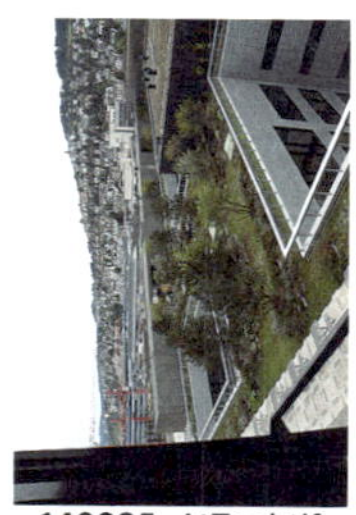
140825_AtToni.tif

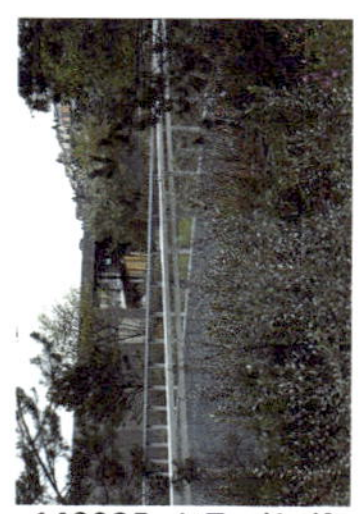
140825_AtToni1.tif

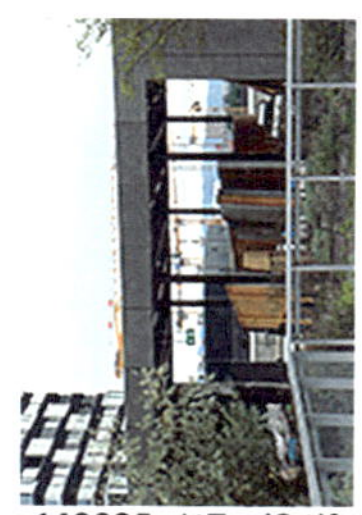
140825_AtToni2.tif

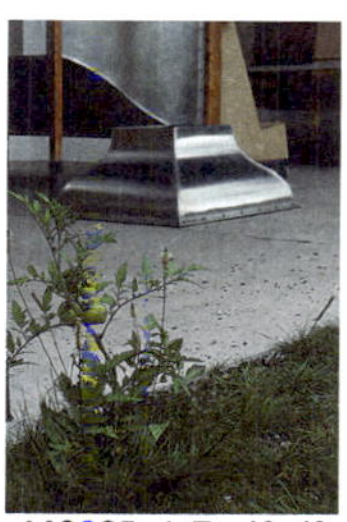
140825_AtToni3.tif

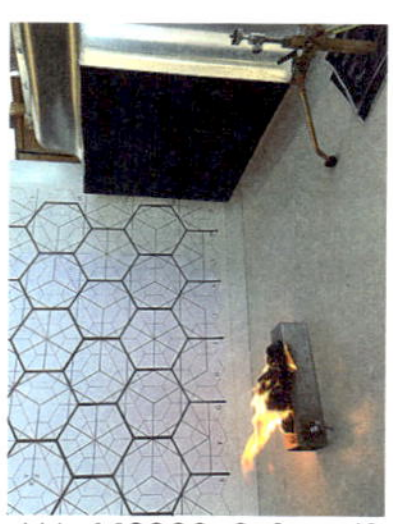
***_140900_Safety.tif

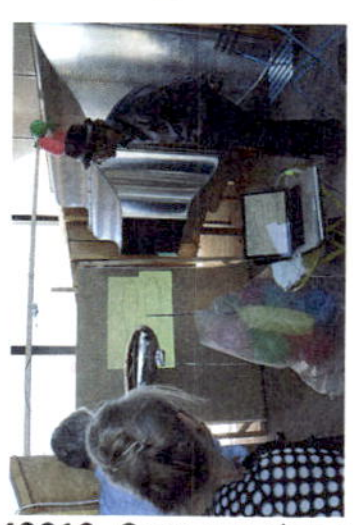
140916_Conversation.tif

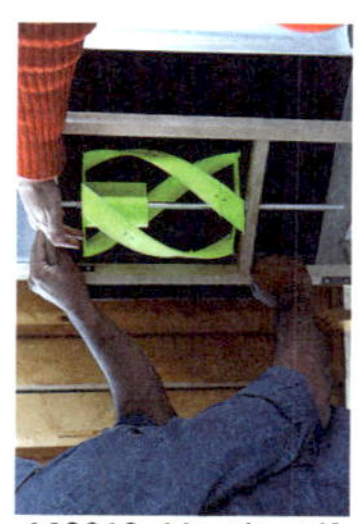
140916_Meeting.tif

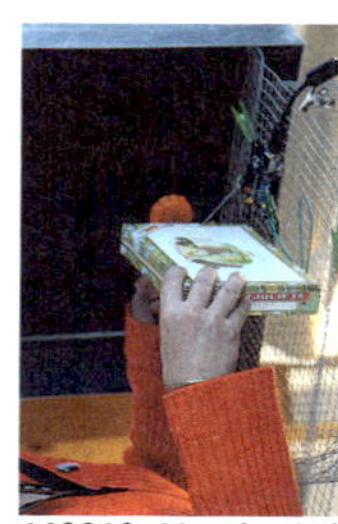
140916_Meeting1.tif

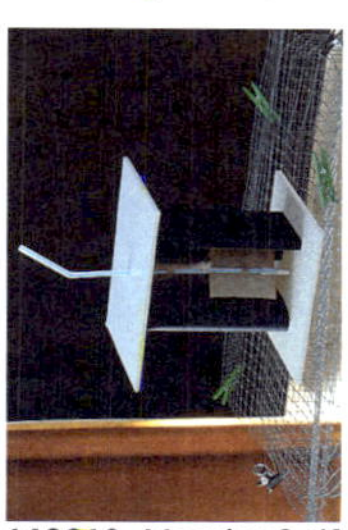
140916_Meeting2.tif

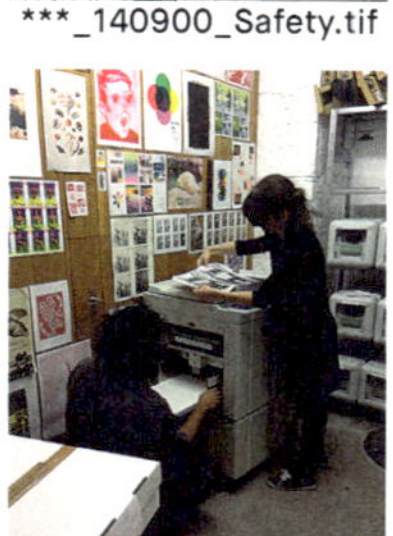
140919_PrintBulletin.tif

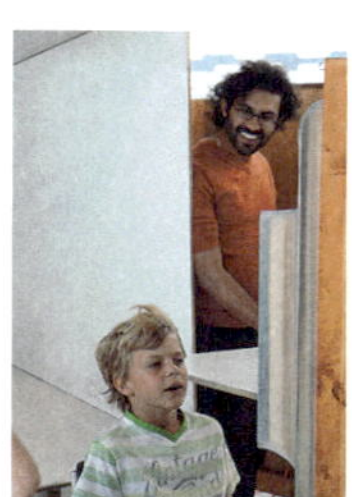
140928_OpenHouse.tif

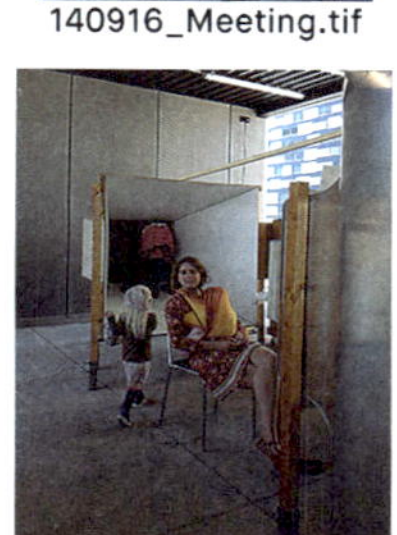
140928_OpenHouse1.tif

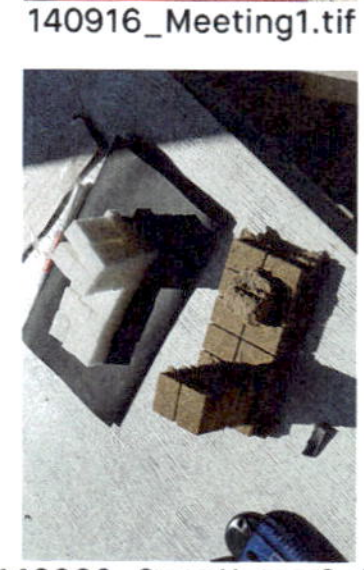
140928_OpenHouse2.tif

Jul 2014
Toni Areal

In July the wind tunnel was taken apart and relocated to the Toni Areal in the western part of Zurich. The newly renovated building houses Zurich's universities of art and of applied sciences and is the workplace of approximately 4000 people. Installing the tunnel and creating a space as useful as the Hafnerstrasse yard at one of Europe's largest art universities would turn out to be more of a challenge than anyone had expected. The Toni Areal was designed to meet all possible safety regulations, and no one seems to have anticipated the impact of this systematic exclusion of individual responsibility: the high level of surveillance, the heavy and mostly locked doors and the speaking lifts all define movements and soundscapes. However, set a bit apart from this, the wind tunnel still asserts the open quality of research (see September 2013).

140701
Team
JH joins the team as senior researcher in the 'Size Matters' project.

140707
Publication
FD's public artwork *Zugabe*, consisting of two models of buildings, is inaugurated at the Brandenburg State Parliament in Potsdam.

140710–140711
Excursion
The team makes a visit to Brussels including Mini-Europe, the Atomium and the Von Karman Institute for Fluid Dynamics.

140722
Construction
The wind tunnel is reassembled at its new location on the roof of the Toni Areal.

Aug 2014
Draught

The wind tunnel was allocated a covered section on the roof of the Toni Areal which was originally designated for a sculptors' workshop. As part of the roof garden the space is both beautiful and closer to the wind than its previous location. This is not always an advantage, because wind tunnels are sensitive to unruly winds from the outside. This circumstance led to an unending series of requests and discussions until a durable solution was found and two sliding gates were installed in 2016 to keep the draught from un-straightening the tunnel's laminar flows.

140807
Excursion
In Askov, Denmark, JH visits the museum of the windmill pioneer Poul La Cour, inventor of the three-winged mill; the museum contains a series of open wind tunnels for testing windmills.

140824
Conversation
Nora Hauswirth, curator, and Harold Schellinx, professor of mathematics and composer from Paris, visit the wind tunnel.

140827
Multiple
KK develops the concept of *designated discomfort areas*, which appears on a sticker published in *Wind Tunnel Bulletin #3* and on the postcard in *Wind Tunnel Bulletin #4* (see November 2014, June 2015, October 2015).

Sep 2014
Safety

On 27 and 28 September the Toni Areal had its Open House, welcoming thousands of visitors into the new premises. The wind tunnel team celebrated with a series of test models where the Toni Areal was exposed to many (im)possible dangers: surprises like downpours of yogurt, attacks by flying pencils and giants eating the building as if it were a chocolate cake. FD said in his introduction: *"Security rules prevent that something happens. Many security rules can ensure that even less happens. And if nothing happens, then nothing really happens. Is that what we want? I am an artist and it is my task to make sure that something does happen."* The wind tunnel offered the visitors a place to think in the wind, which led to many interesting exchanges.

140915–140916
Conversation
Annika Haas, Saskia Frank and Johannes Bennke from the Art and Sciences Workgroup, Berlin, conduct prepared tests in the wind tunnel.

140916
Meeting
The team builds and tests windmills with the tunnel's wind.

140917
Teaching
SW starts teaching the seminar *Models as Media in Film and Science* at the Department of Media Studies of the University of Basel with Stefanie Breuer.

140919
Publication
The group show *The Allegory of the Cave Painting* opens at ExtraCity Kunsthal in Antwerp, where FD exhibits a work inspired by wind streams and HA writes a text for the catalogue (see Selected Publications).

140919
Team
Kasimir is born.

140923
Construction
A temporary wall is added to the outside of the space on the roof to fend off some of the natural winds.

140927–140928
Publication
Toni Areal Open House: ZHdK opens its doors of the new building and welcomes 30 000 visitors.

Johannes Bennke:
The wind tunnel is a straightening, continuous automaton with a randomizer. With it, unsophisticated dabbling becomes a healthy exercise in self-alienation.

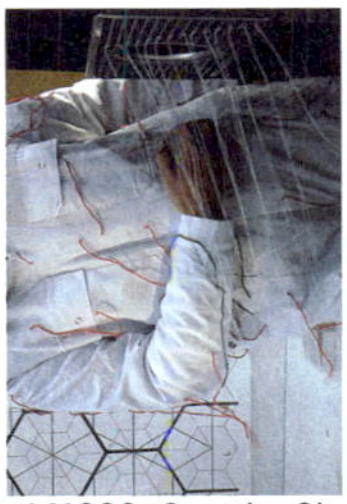
***_141000_CreativeCity.tif

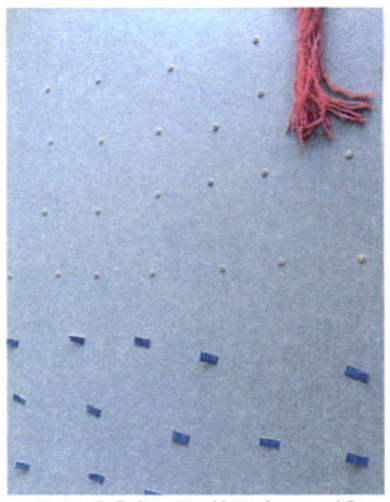
141009_TellTales.tif

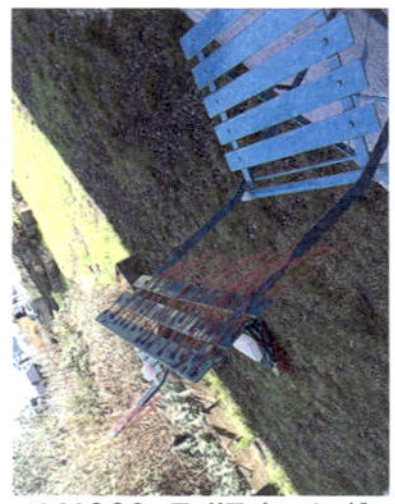
141009_TellTales1.tif

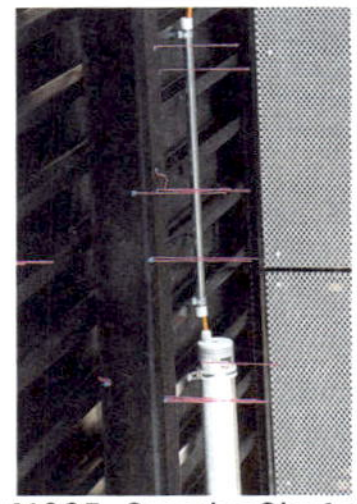
141025_CreativeCity1.tif

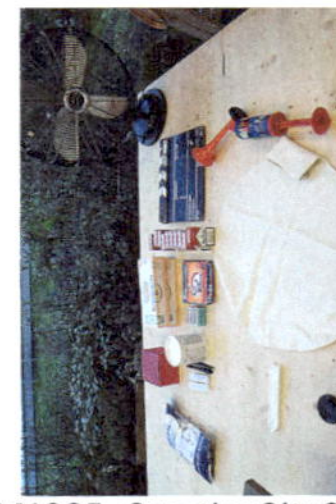
141025_CreativeCity2.tif

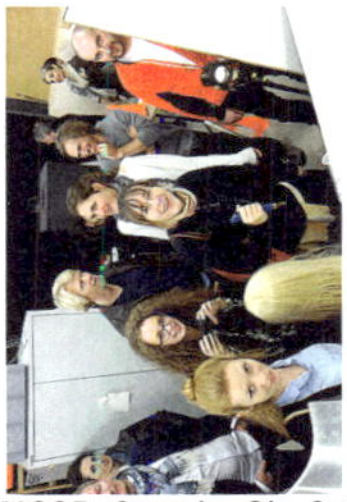
141025_CreativeCity3.tif

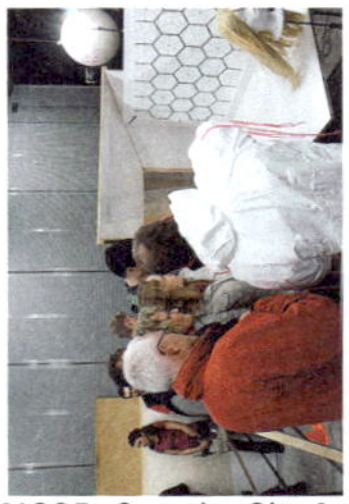
141025_CreativeCity4.tif

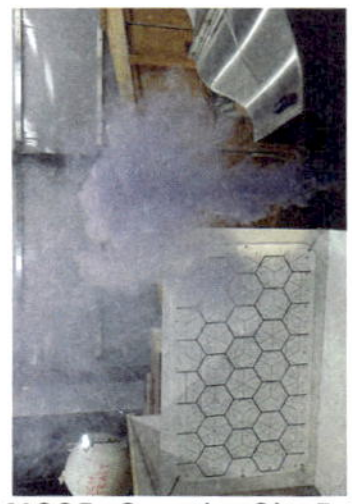
141025_CreativeCity5.tif

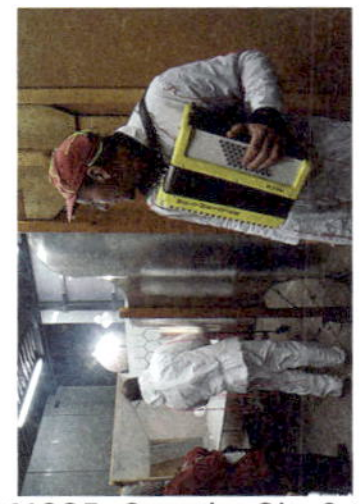
141025_CreativeCity6.tif

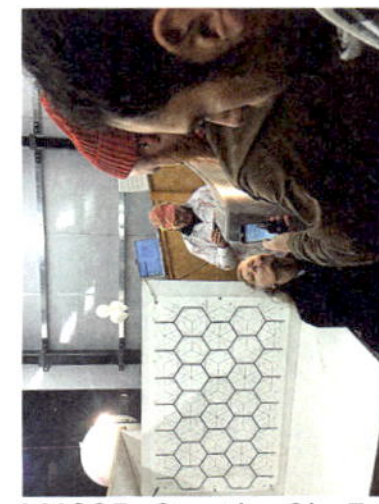
141025_CreativeCity7.tif

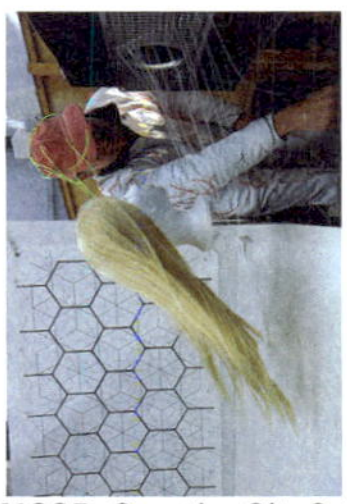
141025_CreativeCity8.tif

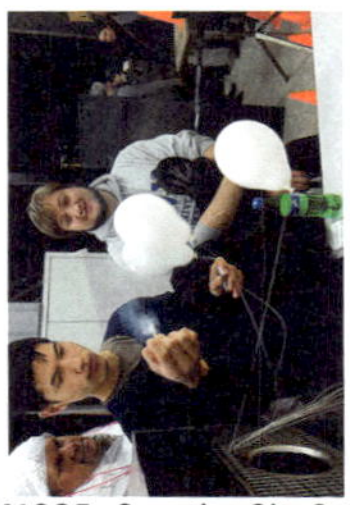
141025_CreativeCity9.tif

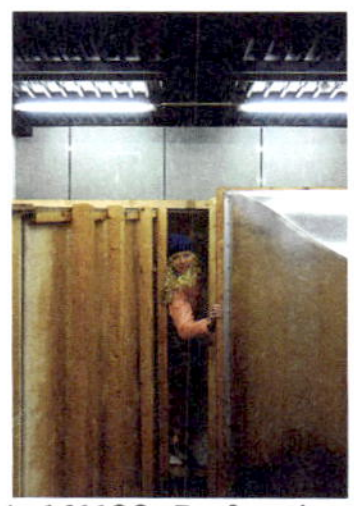
***_141100_Performing.tif

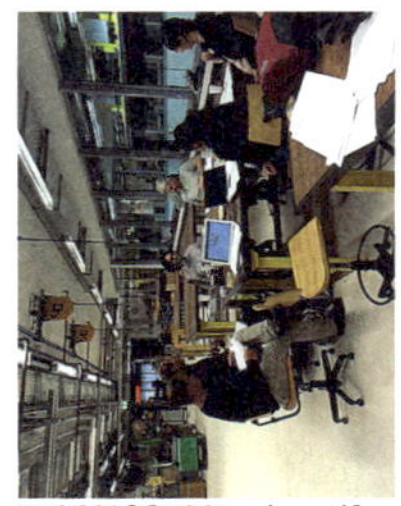
141103_Meeting.tif

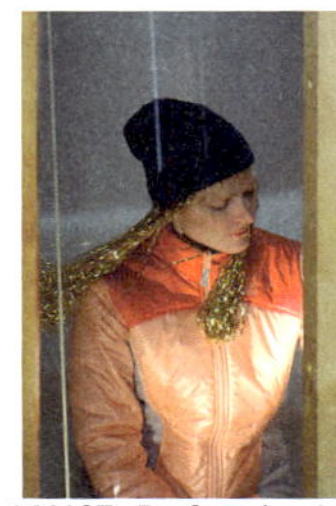
141127_Performing1.tif

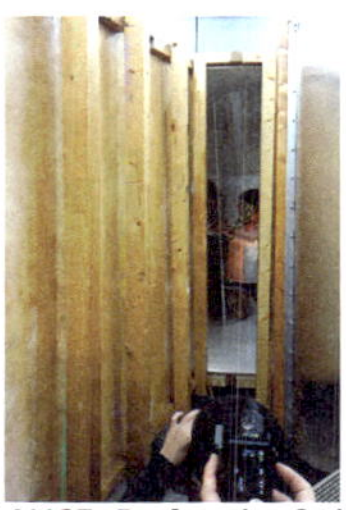
141127_Performing2.tif

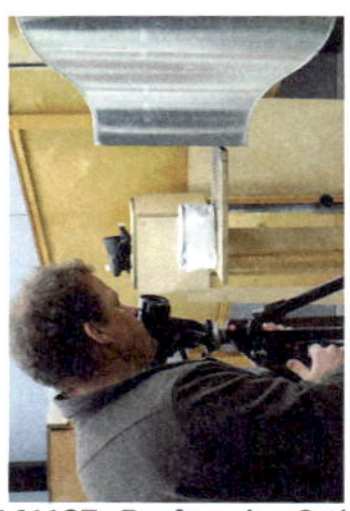
141127_Performing3.tif

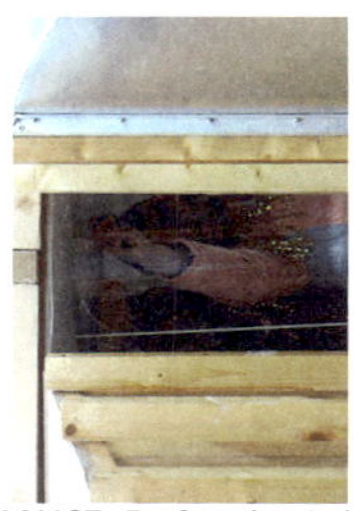
141127_Performing4.tif

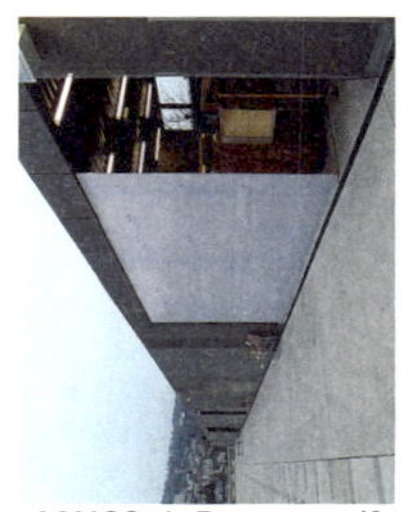
141129_InBetween.tif

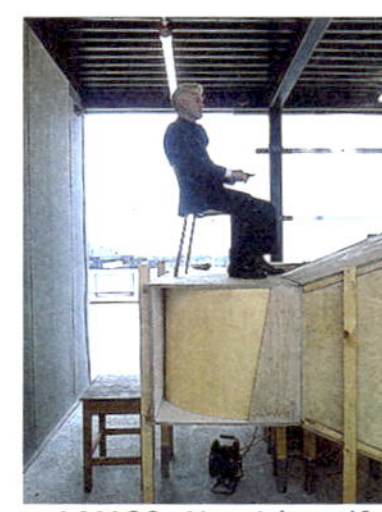
141129_NewIdea.tif

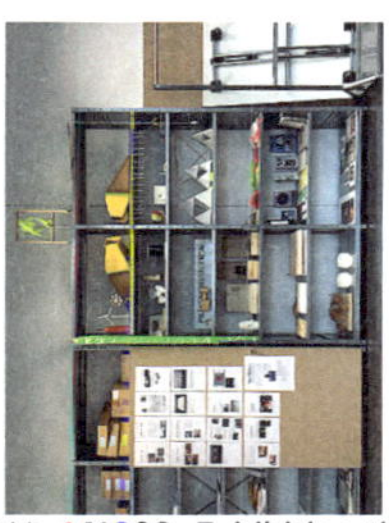
***_141200_Exhibiting.tif

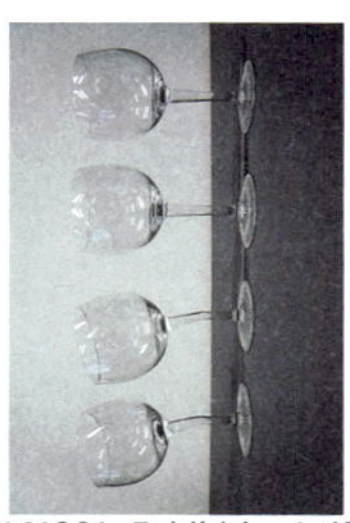
141201_Exhibiting1.tif

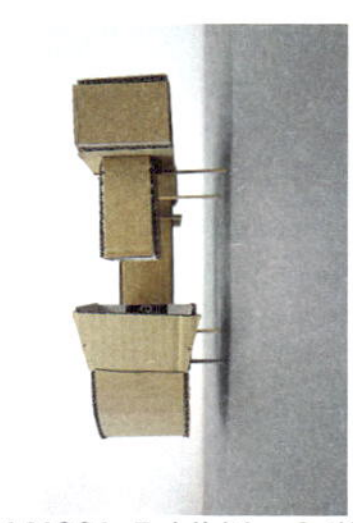
141201_Exhibiting2.tif

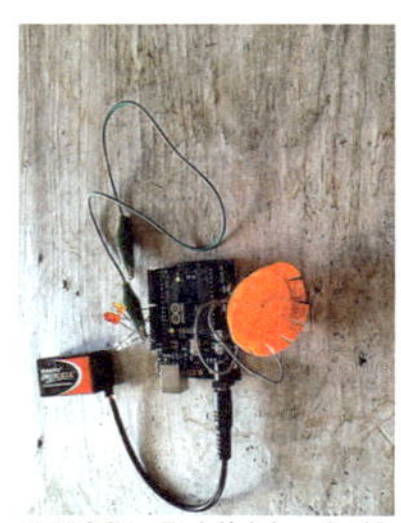
141201_Exhibiting3.tif

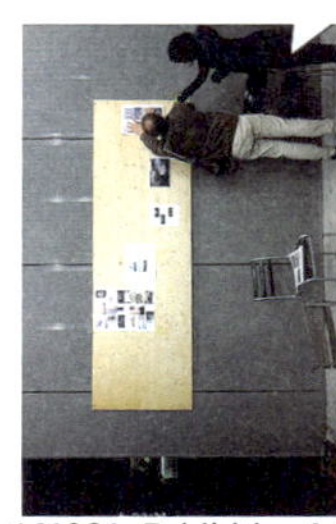
141201_Exhibiting4.tif

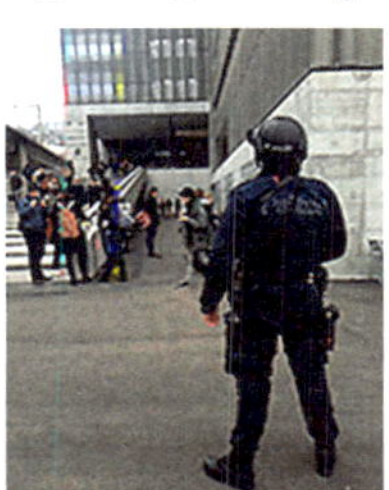
141204_AmokAlarm.tif

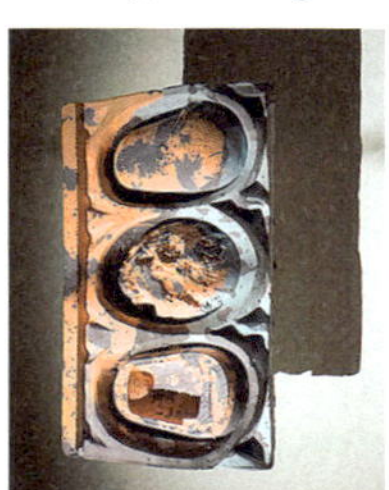
141209_Albatross.tif

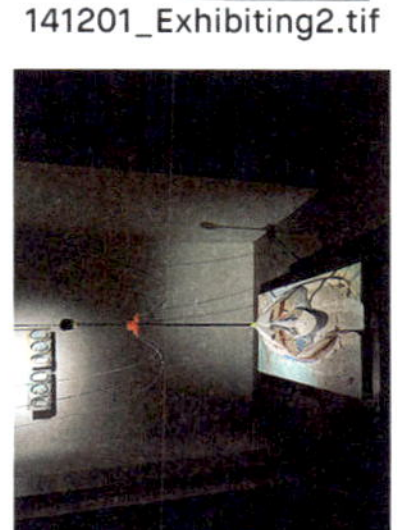
141209_Albatross1.tif

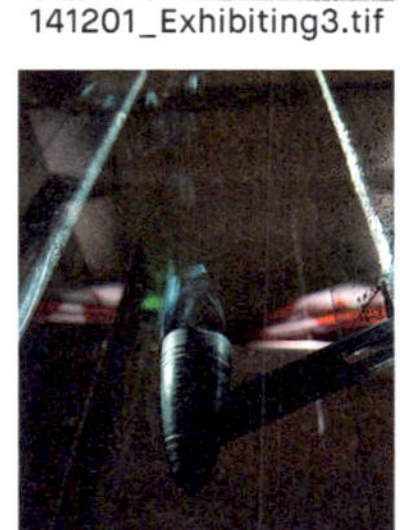
141209_Albatross2.tif

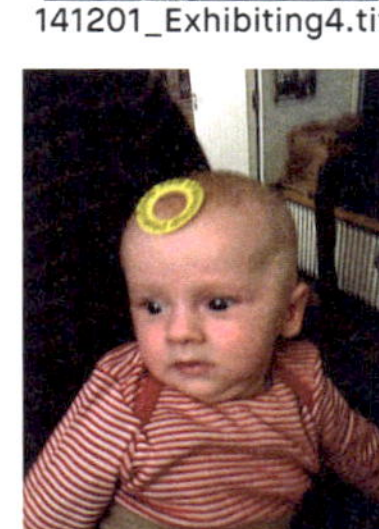
141218_Discomfort.tif

Oct 2014
Alarming

The party 'Creative City' was another invitation to the public of Zurich to get acquainted with the ZHdK, and thousands of people passed by to have a look at the wind tunnel during their tour of the building. Those interested were introduced to the tunnel's gale and gust. Smoke, which is an important part of the tunnel's repertoire, regrettably also turned out to be controversial, as it activated the fire alarms and, with them, the facility management. Our hospitality included Taifun beer and deliciously thin crêpes and attracted visitors. In spite of their responsible containment, the flames of the outdoor oven also added some heat to the interaction. All this, and the sounds of fire, music and discussion created cracking moments of tension and magic.

141009
Multiple
FD brings up the concept of the 'tell-tale', a piece of pink yarn with a super magnet that can be attached to any magnetic ground. The tell-tales are distributed in the Toni Areal, converting the building into a giant wind tunnel.

141021
Meeting
The team members compose personal statements on their relationship to institutions and discuss literature on institutional critique and new institutionalism.

141025
Publication
ZHdK introduces the public to its work in the 'Creative City' event and welcomes 10 000 visitors.

Nov 2014
Performing

The wind tunnel is also a stage and MB, Magnhild Fossum and Alice McCabe collaborated on testing different items moving through the interior space and celebrating the medieval tradition of a monthly fools' day. The moving glitter was caught on camera. The action was barely visible through the small slit that let only a little light into the curve. Soon the curve would be expanded to create a larger test section and, with it, an interior stage that could be seen from the outside through a transparent panel (see January 2015).

141103
Meeting
Team members' short texts about central experiences, themes and the important outcomes of the project are discussed in the meeting.

141110
Bulletin
The Wind Tunnel Bulletin #3 is published.

141110
Multiple
The yellow sticker *designated discomfort area* (see August 2014, June 2015, October 2015) is found in the *Wind Tunnel Bulletin #3*.

141121
Conversation
Gabu Heindl and Drehli Robnik give a lecture and present a screening of their film *Mock-Ups in Close-Up: Architectural Models in Film, 1919–2012*.

141127
Conversation
Performance with MB, KK, Magnhild Fossum and Alice McCabe.

Dec 2014
Exhibiting

The wind tunnel was one point of interaction between the Research Focus in Transdisciplinarity and the MA Transdisciplinary Studies. Furthermore, in the Hafnerstrasse and later in the Toni Areal a set of shelves were installed in the MA studio. These formed a small interactive museum for the 'Size Matters' project. Here one could find publications from the project and objects used or made for tests in the wind tunnel. Over time this established a conversation-through-materials between the team and the students, as the objects and texts were organized and reorganized.

141204–150104
Publication
Exhibition *A is for Albatross* by HA at Museum Bärengasse, Zurich.

141209
Meeting
The group is welcomed into the old premises at Hafnerstrasse and visits the mosque and community centre there. Later in the day they visit and have a discussion at HA's exhibition at the Museum Bärengasse.

141219
Publication
Premiere of Piet Esch's film *Hothouse Uerikon* at Xenix Cinema, Zurich.

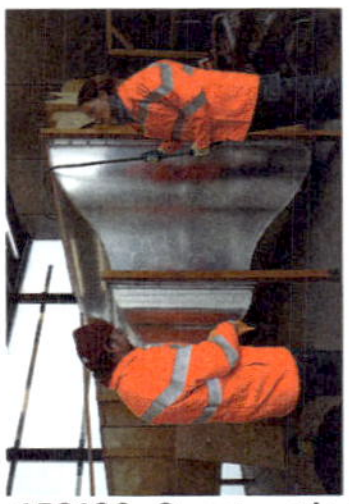
***_150100_Construction.tif

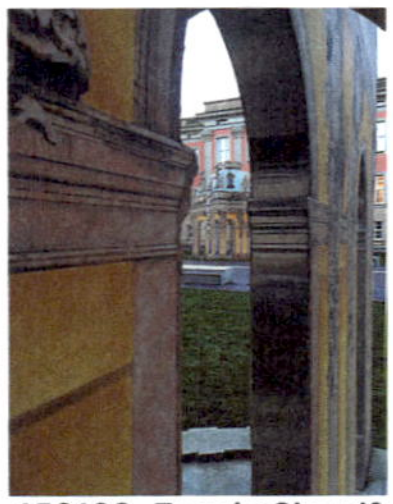
150102_ZugabeSitu.tif

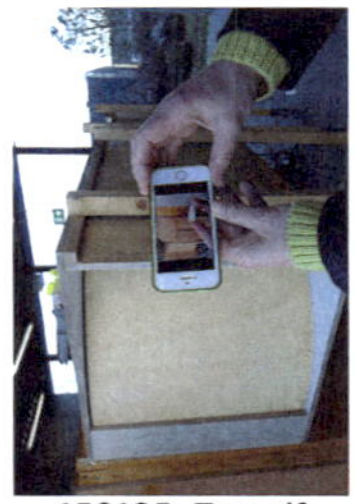
150105_Test.tif

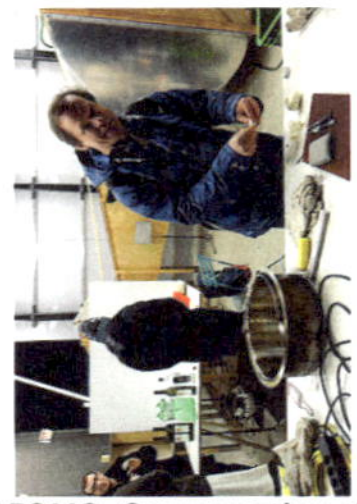
150119_Conversation.tif

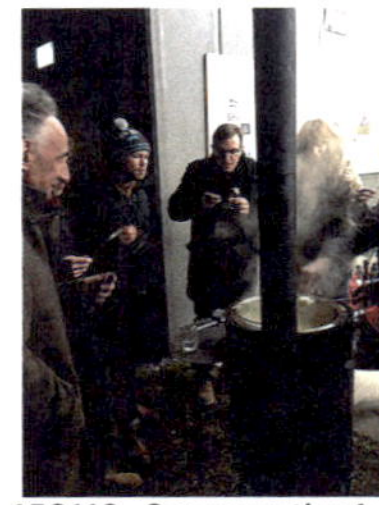
150119_Conversation1.tif

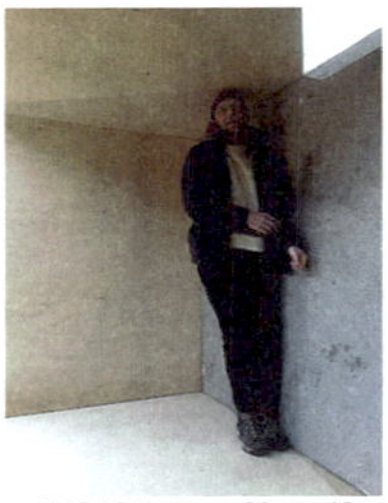
150125_TestSize.tif

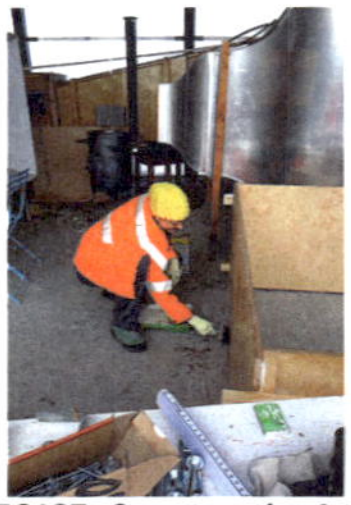
150127_Construction1.tif

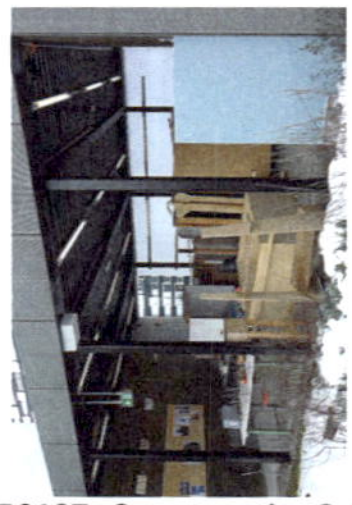
150127_Construction2.tif

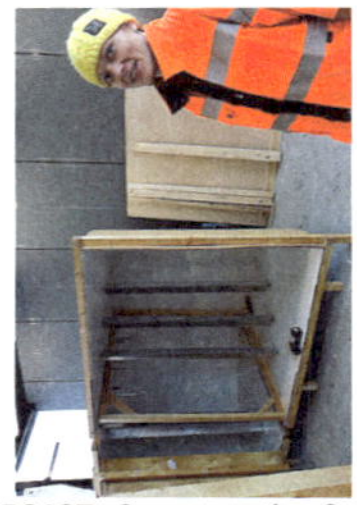
150127_Construction3.tif

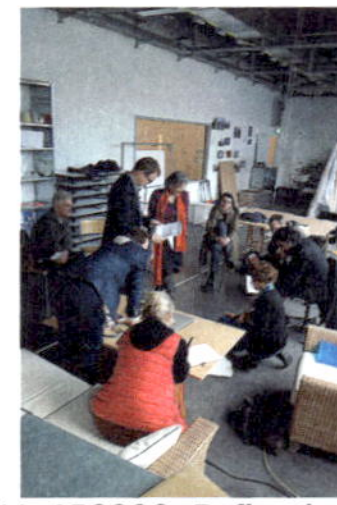
***_150200_Reflection.tif

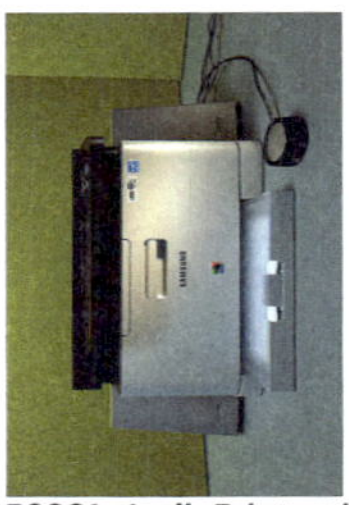
150201_AudioPrinter.tif

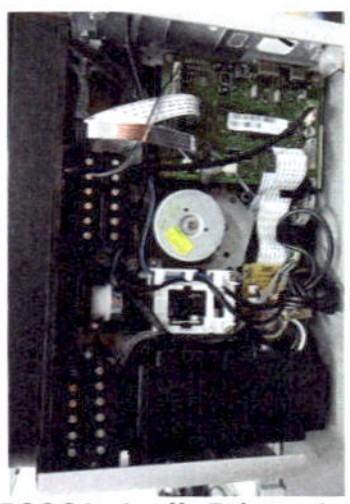
150201_AudioPrinter1.tif

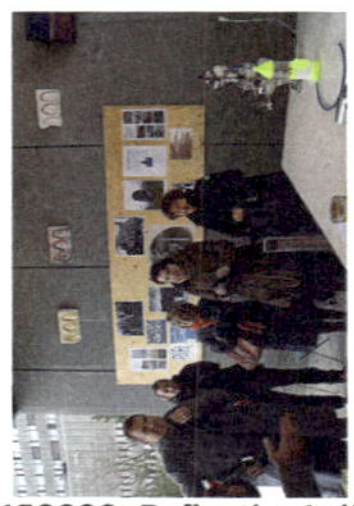
150202_Reflection1.tif

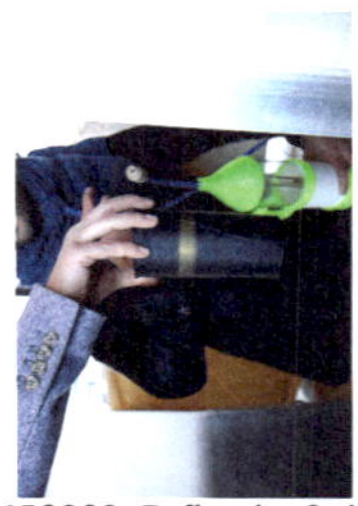
150202_Reflection2.tif

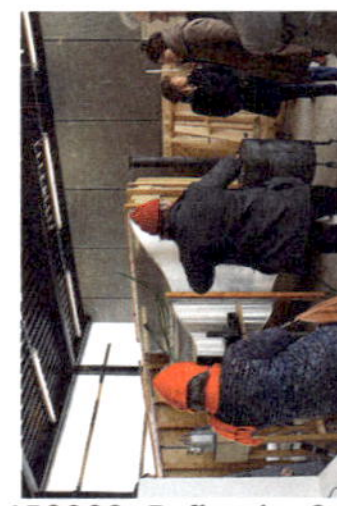
150202_Reflection3.tif

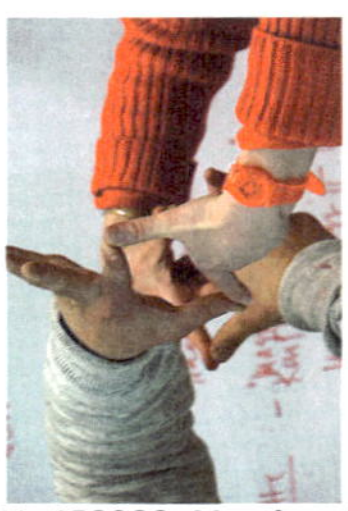
***_150300_Meeting.tif

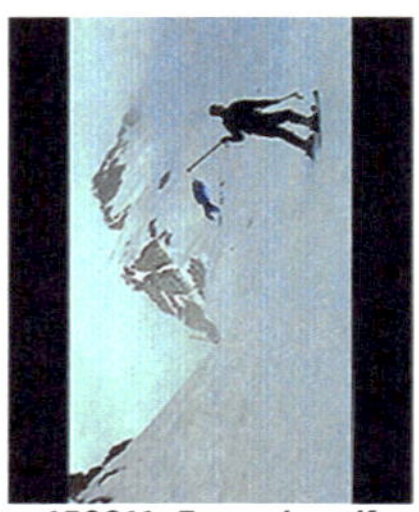
150311_Excursion.tif

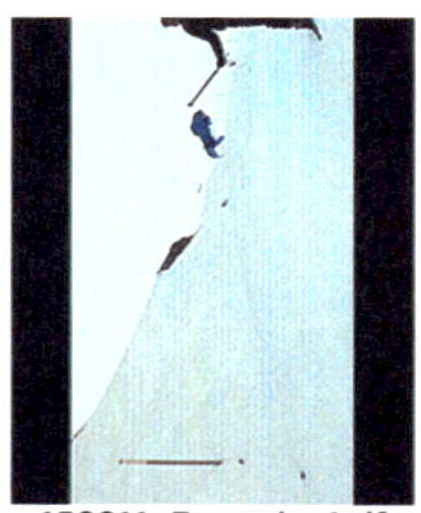
150311_Excursion1.tif

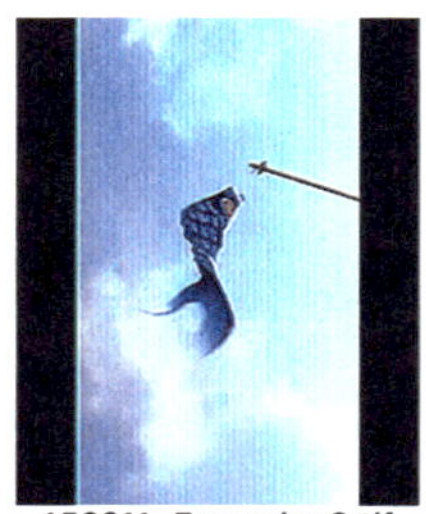
150311_Excursion2.tif

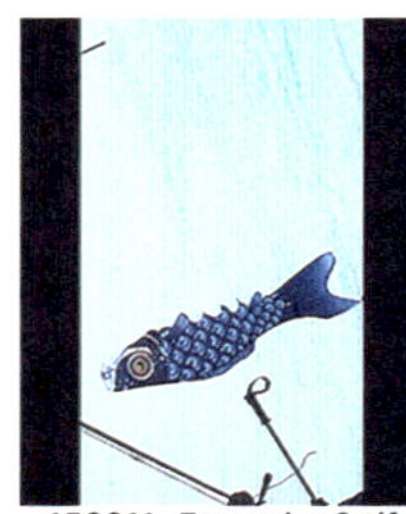
150311_Excursion3.tif

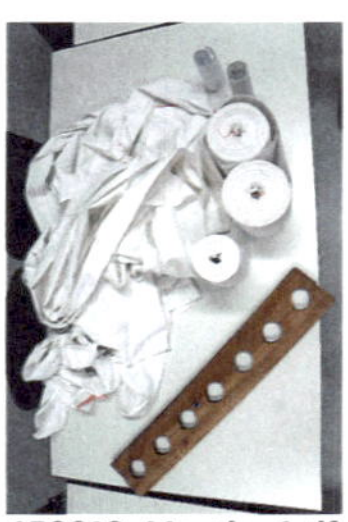
150312_Meeting1.tif

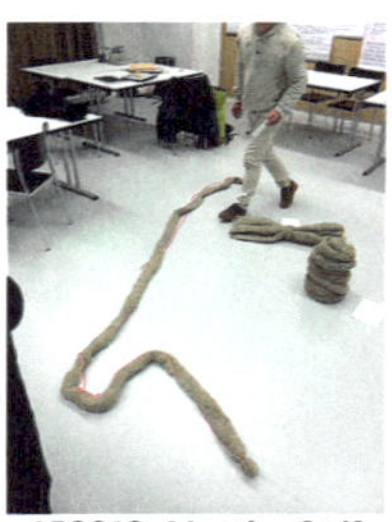
150312_Meeting2.tif

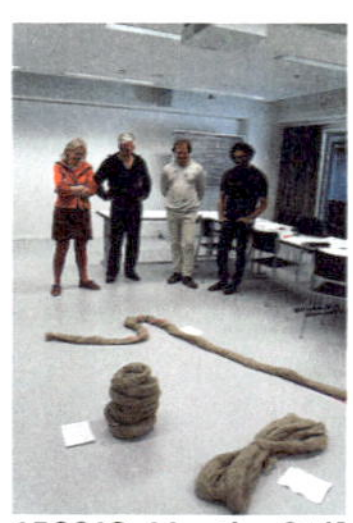
150312_Meeting3.tif

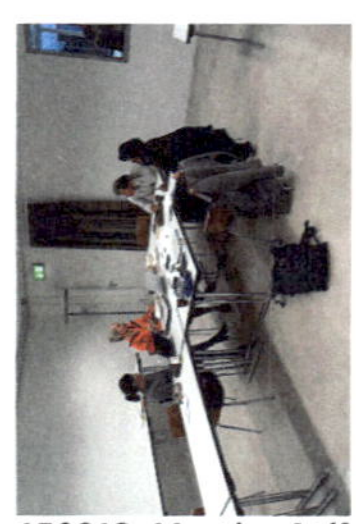
150312_Meeting4.tif

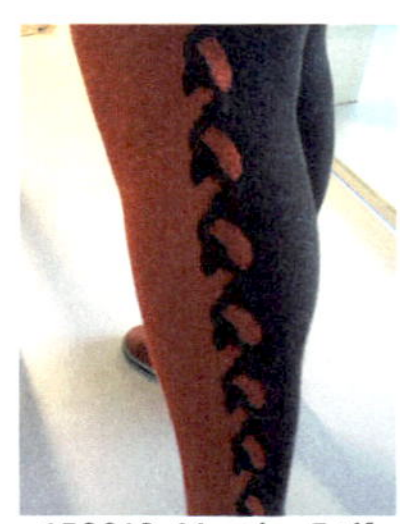
150312_Meeting5.tif

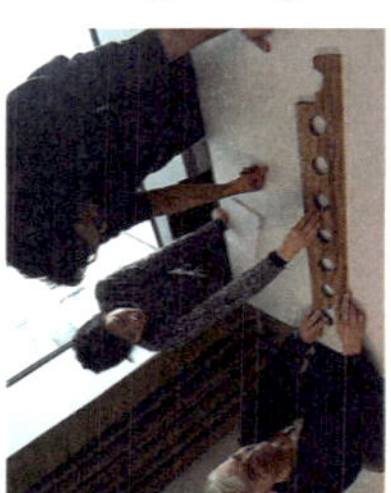
150312_Meeting6.tif

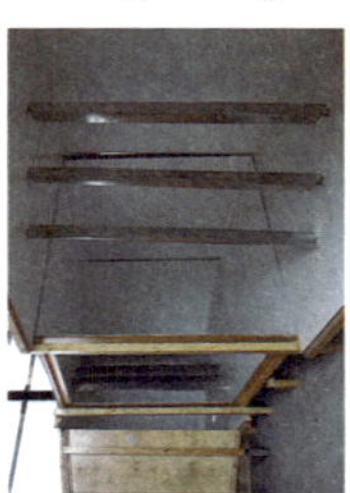
150326_Construction.tif

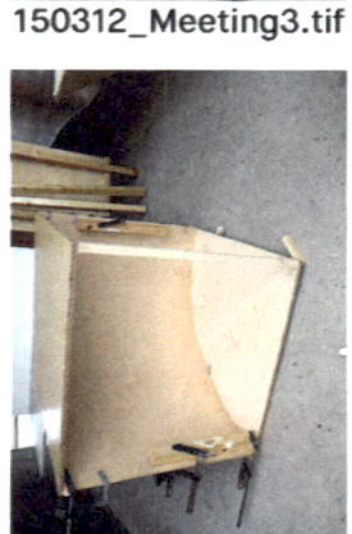
150326_Construction1.tif

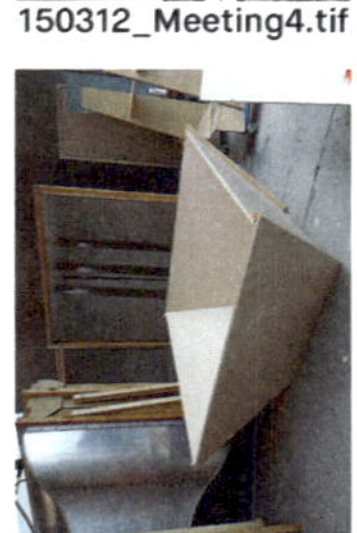
150326_Construction2.tif

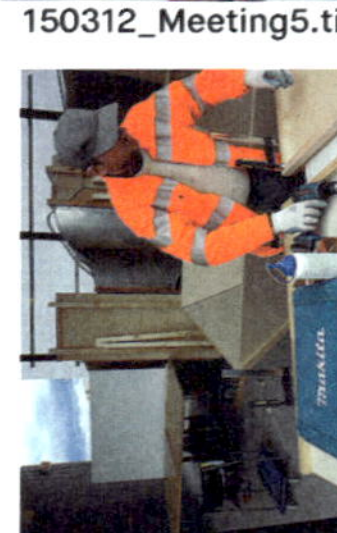
150326_Construction3.tif

Jan 2015
Uniforms

The wind tunnel was about to move on to its next phase through the addition of the second test section and an expanded curve. The work was carried out by KK and Magnhild Fossum; both wore fluorescent work jackets and sturdy trousers. The uniforms were part of a strategy to build trust in the serious endeavour of artistic research. KK: *"The professional outfits supported and expressed the clear goal of the workers: the construction of the tunnel. Moreover, moving around the Toni Areal in the designated uniforms made the hands-on work, the engineering skills, as well as the design and craftsmanship unquestionable."*

150112
Meeting
An invitation for an exhibition at the Wind Tunnel Adlershof in Berlin is discussed and conceptualized, but finally turned down.

150119–150120
Conversation
Luc Döbereiner, composer and artistic researcher from Berlin, gives a talk on his work and his PhD thesis *The Concept of Compositional Model* (University of Music and Performing Arts Graz).

150120
Team
Mio is born.

150127
Construction
Work on the expansion of certain tunnel sections and a new transparent entry is under way.

Feb 2015
Reflection

In February the Reflection Group met for the first of four encounters. Six high-ranking professionals – Jacqueline Burckhardt, Corina Caduff, Christoph Hoffmann, Dieter Mersch, Isabel Mundry and Jan Svenungsson – were invited to exchange ideas and reflect on the work with the wind tunnel. (Some of the results can be read in this book.) In the first meeting the members of the group were introduced to the Toni Areal: from the workshops in the cellar to the concert halls and the studios on the seventh floor. They also listened to elevator pitches of the individual team members and met the wind tunnel. Being on the roof was a cold affair in February and after freezing in the open air, the guests spent some hours getting acquainted with each other's artistic and scientific work in the studio of the MA Transdisciplinary Studies.

150201
Multiple
In the new building no individual printers are allowed. So KK develops the *Audio Printer* by adding a sonification unit and a pair of speakers to an inkjet printer. Now it's art and permitted to be both reflective and useful.

150202
Conversation
The Reflection Group meets for the first time and is introduced to the project.

150224
Conversation
Christina Kubisch, sound artist and professor from Berlin, visits the wind tunnel.

Mar 2015
Fahrkunst

The meeting in March scaled down and accelerated the team's working method of *Fahrkunst* (cf. FD: Fahrkunst, pp 103–105). Two groups were formed and, as in a game of Chinese whispers, exchanged an impulse either in the form of a text or in a visual form. An answer in the opposite format was required. Hence the response to a text would be a visual expression and vice versa. Alternating in short rounds of only 30 minutes this created a fast flow of texts and images, of responses and ideas circumscribing the question of the exchange. As MB put it: *"These small acts also reflected their quality of recycling, it was and is about using material from one context and scaling it into a new meaning."*

150311
Excursion
Skiing and sledge tours at Sörenberg, in central Switzerland, again experiencing and studying wind by moving ourselves.

150312
Meeting
The team exercises and reassesses the concept of *Fahrkunst*.

150326
Construction
The expanded curve is almost complete and is tested as a stage.

Luc Döbereiner:
In the wind tunnel, you find yourself within a turbulence between the abstract and the concrete.

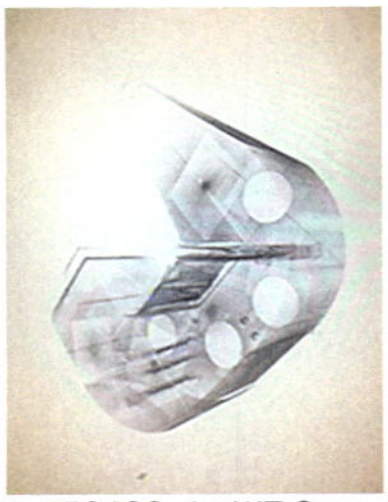
***_150400_1stWT-Scan.tif

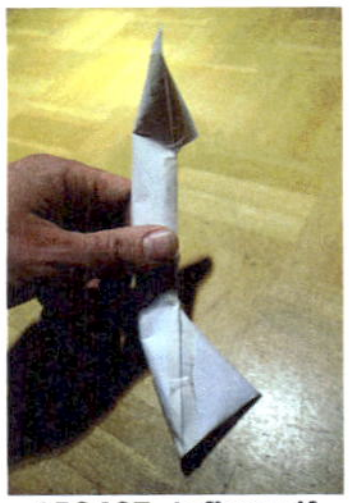
150407_Inflate.tif

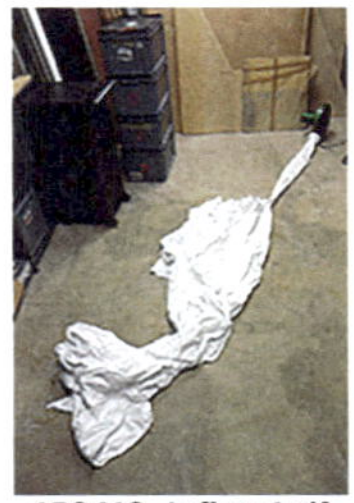
150418_Inflate1.tif

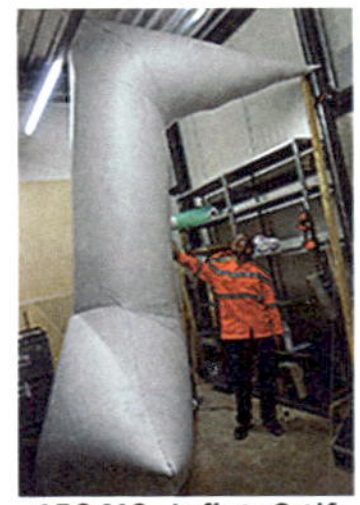
150418_Inflate2.tif

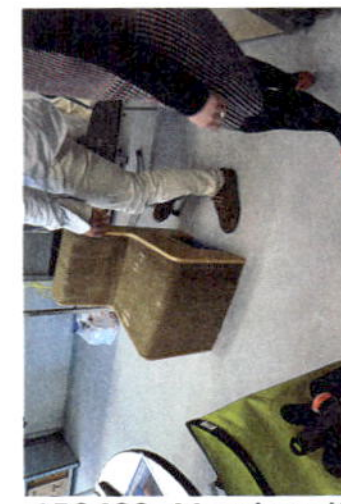
150423_Meeting.tif

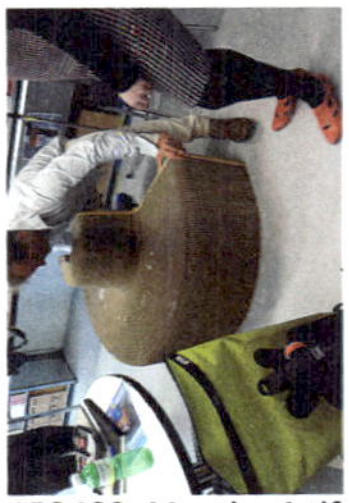
150423_Meeting1.tif

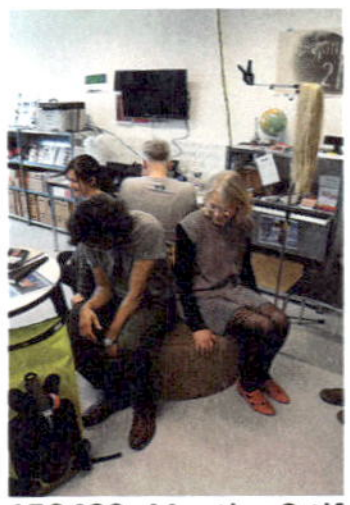
150423_Meeting2.tif

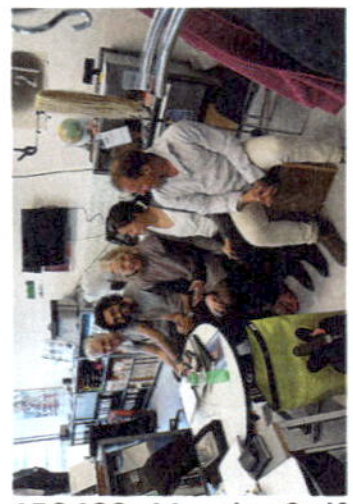
150423_Meeting3.tif

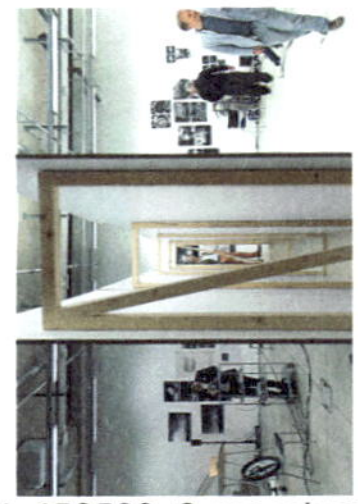
***_150500_Symposium.tif

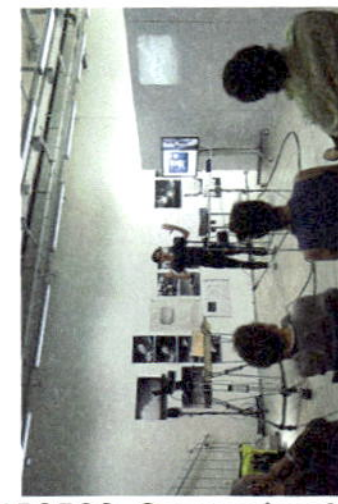
150508_Symposium1.tif

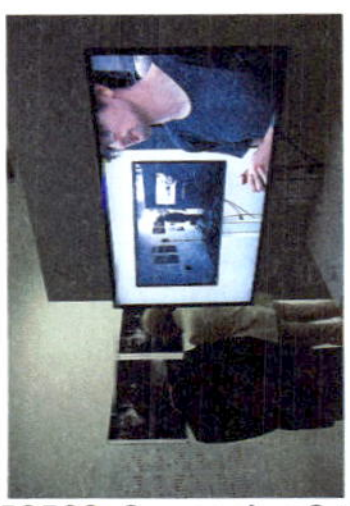
150508_Symposium2.tif

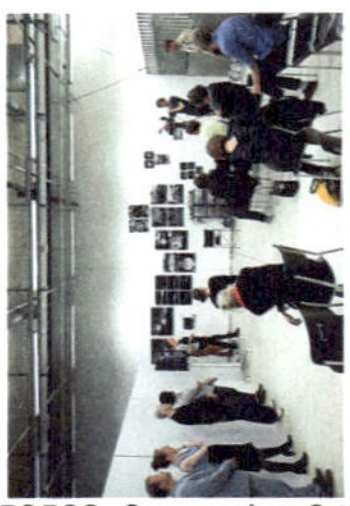
150508_Symposium3.tif

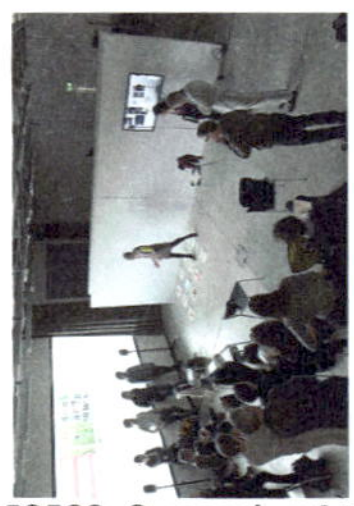
150509_Symposium4.tif

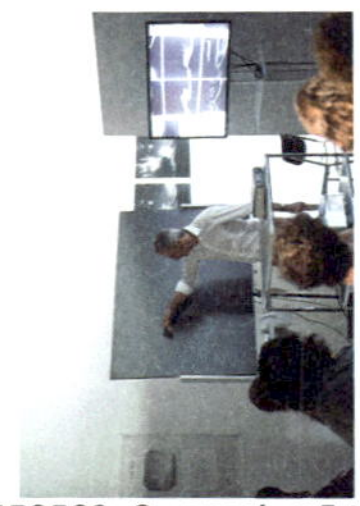
150509_Symposium5.tif

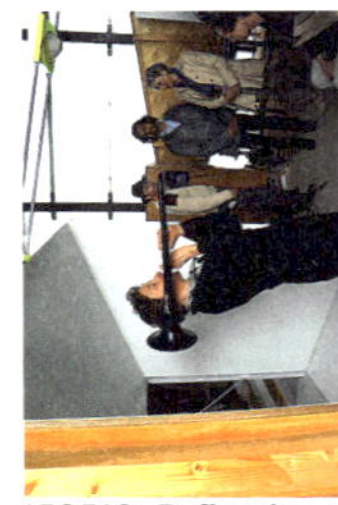
150519_Reflection.tif

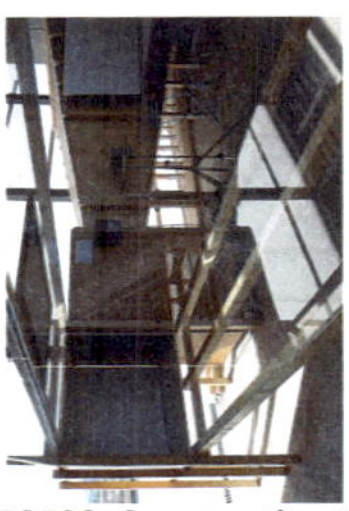
150528_Construction.tif

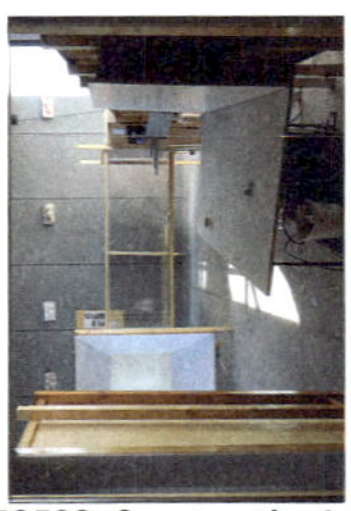
150528_Construction1.tif

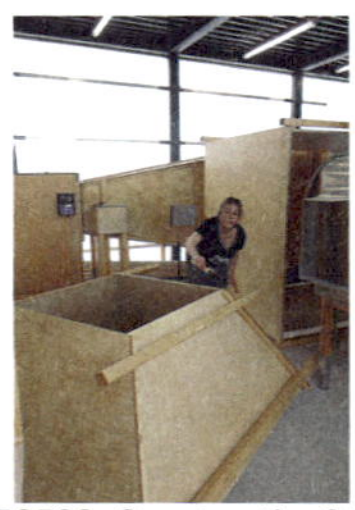
150528_Construction2.tif

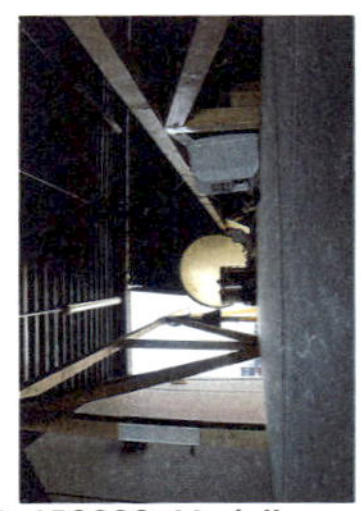
***_150600_Modellapero.tif

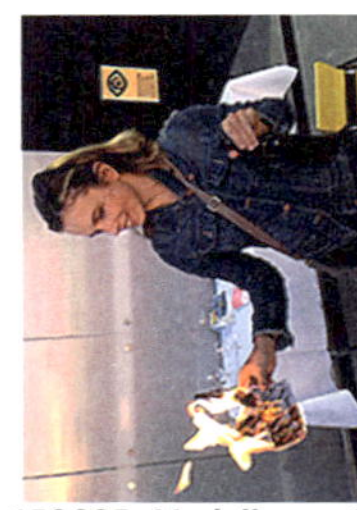
150625_Modellapero1.tif

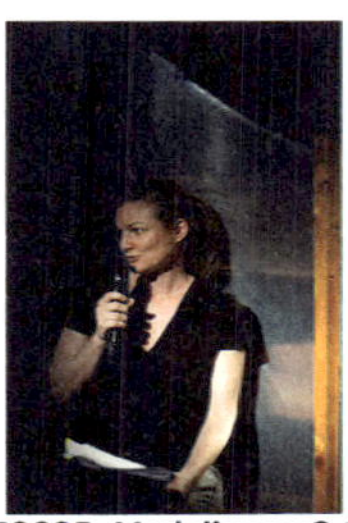
150625_Modellapero2.tif

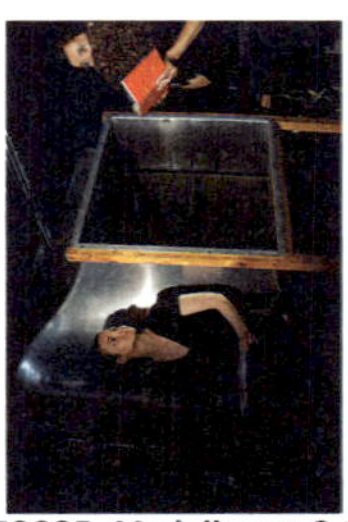
150625_Modellapero3.tif

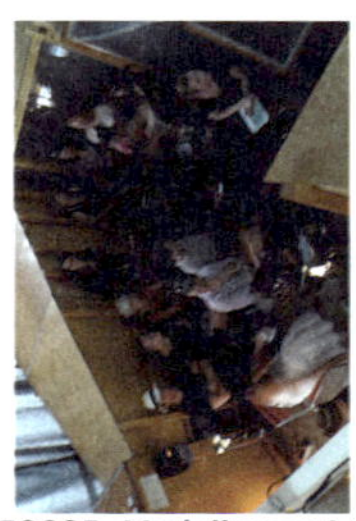
150625_Modellapero4.tif

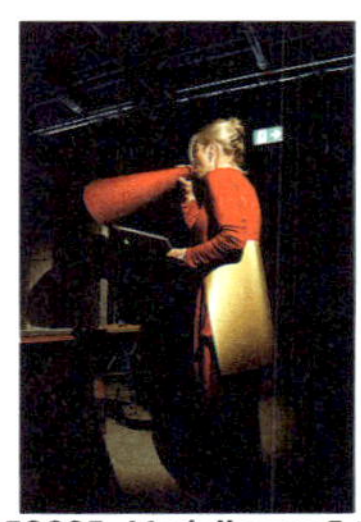
150625_Modellapero5.tif

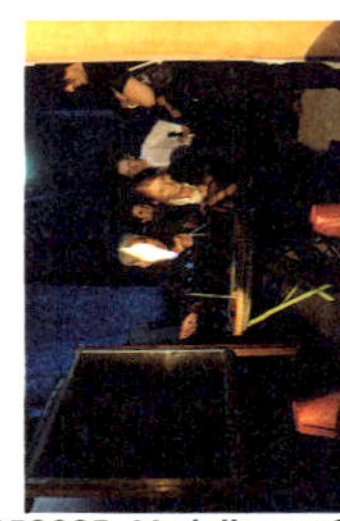
150625_Modellapero6.tif

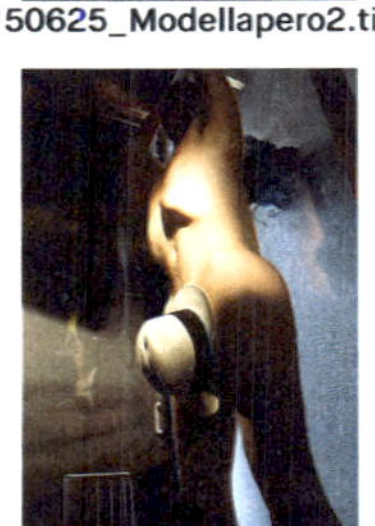
150625_Modellapero7.tif

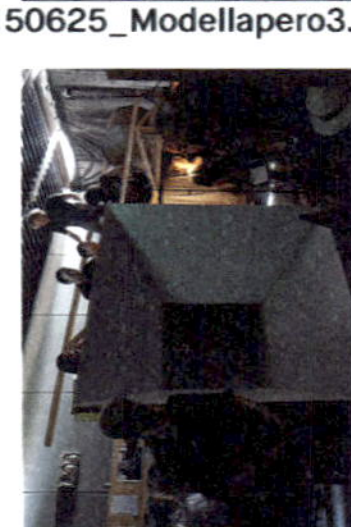
150625_Modellapero8.tif

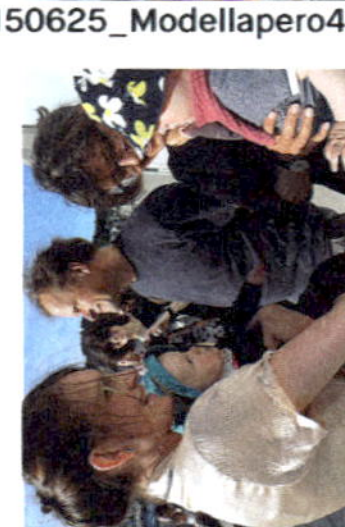
150625_Modellapero9.tif

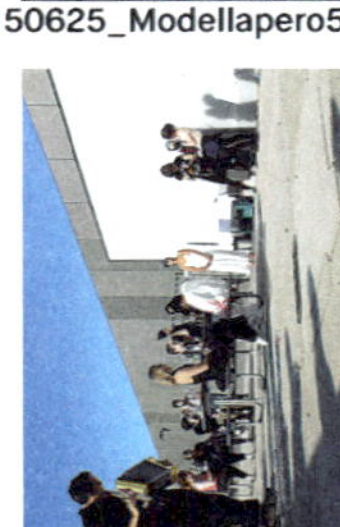
150625_Modellapero10.tif

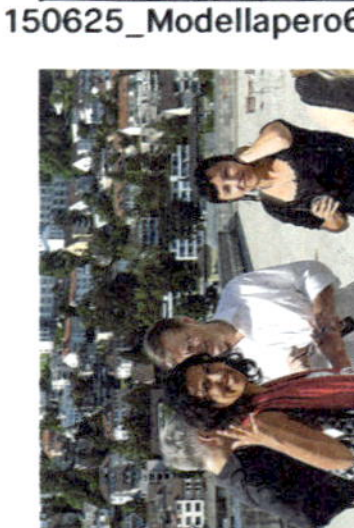
150625_Modellapero11.tif

Apr 2015
The Void

By scanning the wind tunnel, Jonathan Banz revealed an image of the tunnel from the inside. The resulting images seem to finally unite the tunnel with its transparent partner, the wind, and defy gravity (see October 2015).

150414
Conversation
Jonathan Banz of the Swiss Federal Institute of Technology Zurich makes a digital scan of the inside of the wind tunnel.

150423
Meeting
The team reads texts on the concept of 'boundary objects', works on a vocabulary of scale and develops small actions about scale using exercises László Moholy-Nagy introduced at the Institute of Design in Chicago.

May 2015
Palaver

The team organized a two-day exchange on scale. The symposium was set in the *Palaver* format (a concept developed by FD and Eran Schaerf 2008), where the space is divided into two areas by a double *paravent*. One side was intended for presentations and the other for discussion. The spaces were separated optically but not acoustically. A screen was mounted on either side of the wall, and cameras transferred the activity from one side to the other. The symposium was a combination of scientific approaches to scale and artistic, imaginative ways of pursuing and playing with the art of zooming out, expanding, enlarging and reducing or focusing.

150508–150509
Publication
Symposium *too big to scale* with Olivier Chazot, Von Karman Institute for Fluid Dynamics, Brussels; Andrew Fisher, Goldsmiths, University of London; Simon Grand, University of St Gallen; Juliet Koss, Columbia University, New York/Scripps College, California; Adrien Lucca, artist, Antwerp; Steeve Sabatto, Ecole des Hautes Etudes en Sciences Sociales, Paris; Frank Schweitzer, ETH Zurich; Helmut Völter, Bureau Spector, Leipzig; Christoph Weckerle, ZHdK; and Reinhard Wendler, Institute for Art History, Florence.

150519
Conversation
The Reflection Group meets for the second time and tests the small-scale *Fahrkunst* (see March 2015).

150522
Conversation
James Elkins, Professor at the Art Institute of Chicago, visits the wind tunnel.

150528
Construction
The combined Göttingen and Eiffel format is complete, with the large transparent test section in front of the engine.

150531
Publication
FD is invited for a keynote lecture at the Research Pavilion of the Venice Biennial. Two years later, in the summer of 2017, the wind tunnel will be exhibited there.

James Elkins:
It was a wonderful surprise to see the wind tunnel: every art university should have one! As an allegory for the erosive force of the art market, of reputation and time, of the inevitable disappearance of all things that are made.

Jun 2015
Varieté Voltaire

Before Zurich was flooded with jubilee events, the team celebrated the DADA centenary from 1916 by turning the wind tunnel into a stage. Visitors to the *Modellapéro* were offered the chance to see a variety programme. The audience was placed in the centre of the wind tunnel as well as on top of it. The team performed windy acts that referenced and questioned the DADA masters. There was an attempt to leave the sacred aura and respond to present anxieties. In the transitions between acts, quotes from the ZHdK safety regulations were read by Delphine Chapuis Schmitz. MB struck a big drum, thereby keeping the performance in rhythm.

150601
Team
Delphine Chapuis Schmitz, artist and assistant at MA Transdisciplinary Studies, joins the team for two months working on the architecture of Ludwig Wittgenstein and its relation to scale.

150603
Construction
A tribune above the newly constructed extension to the Eiffel test section built with the support of the visiting artist Johannes Michler from Kiel creates space for additional seating allowing a good view of the entire installation.

150615–150618
Publication
FD and KK conduct a workshop on wind, scale and utopia at the University of Arts of Cuba as part of the Havana Biennial.

150624
Meeting
Coordination and exchange on the different ongoing publications.

150625
Modellapéro
Varieté Voltaire is performed and, afterwards, spontaneous music sessions occur around the wind tunnel.

150625
Bulletin
Wind Tunnel Bulletin #4 on the theme of *Fahrkunst* is launched.

150625
Multiple
A postcard referring to the Toni Areal as a *designated discomfort area* is included in the Bulletin (see August 2014, November 2014 and October 2015).

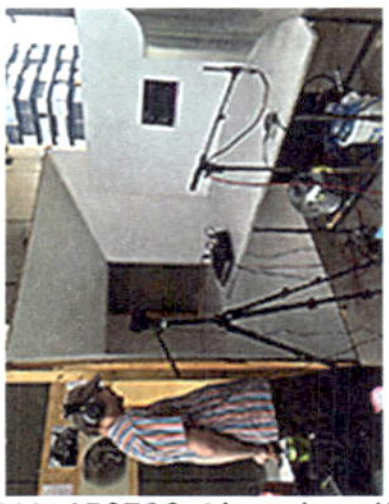
***_150700_Listening.tif

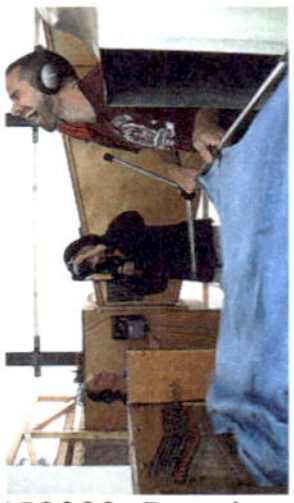
***_150800_Experiment.tif

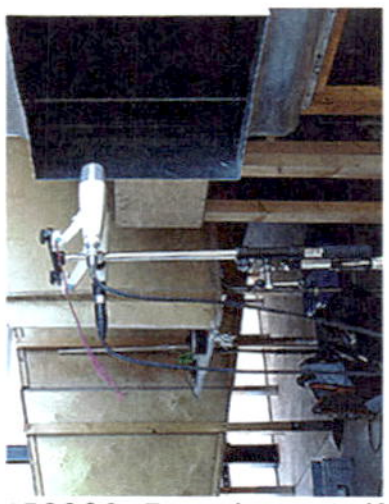
150803_Experiment1.tif

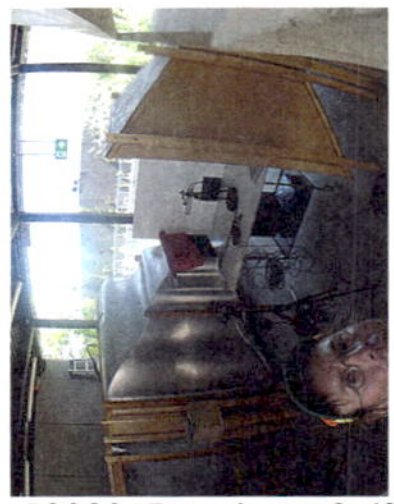
150803_Experiment2.tif

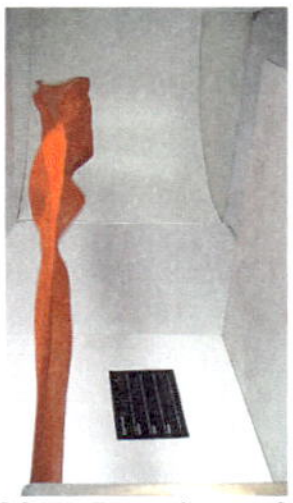
150811_Experiment3.tif

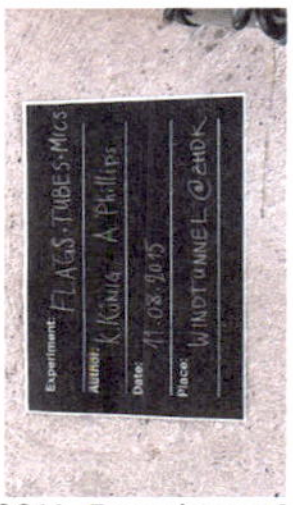
150811_Experiment4.tif

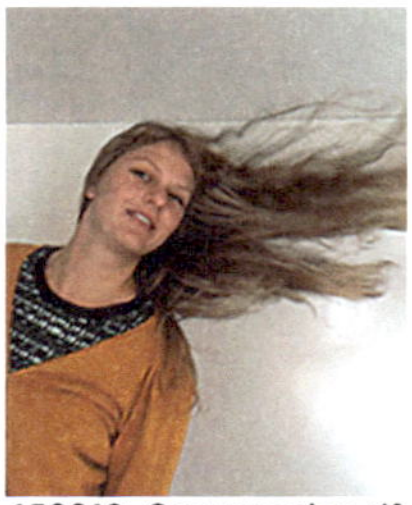
150819_Conversation.tif

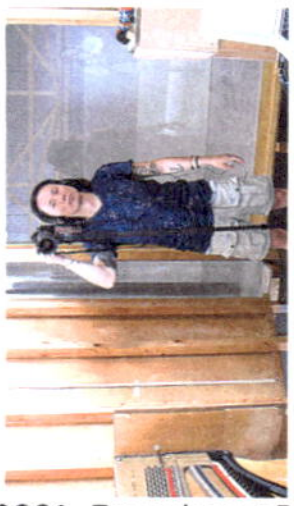
150831_Experiment5.tif

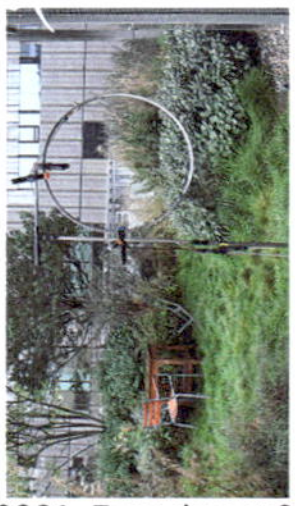
150831_Experiment6.tif

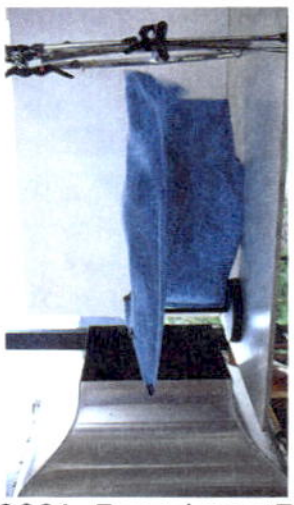
150831_Experiment7.tif

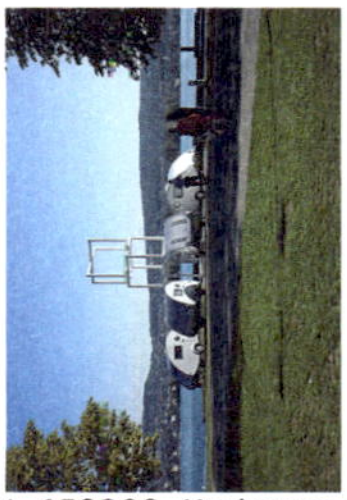
***_150900_Hothouse.tif

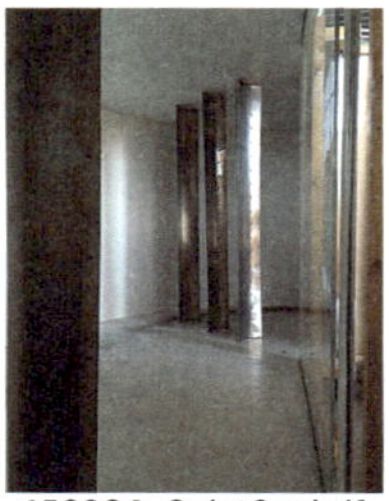
150904_QuietCool.tif

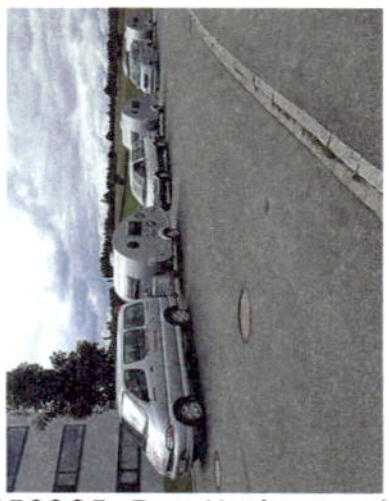
150905_PrepHothouse.tif

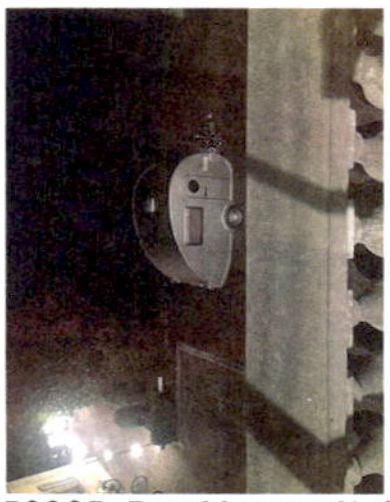
150905_PrepMuseumN.tif

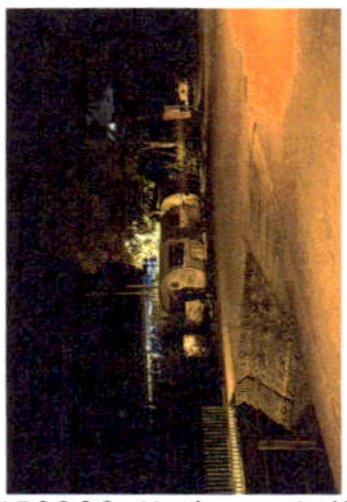
150906_Hothouse1.tif

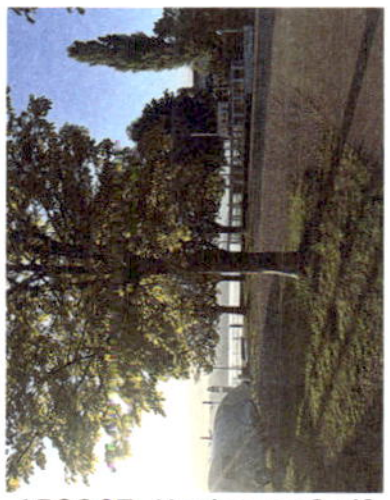
150907_Hothouse2.tif

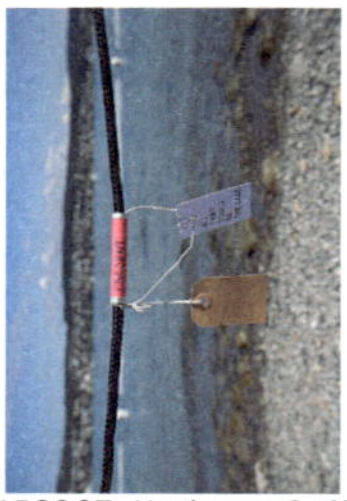
150907_Hothouse3.tif

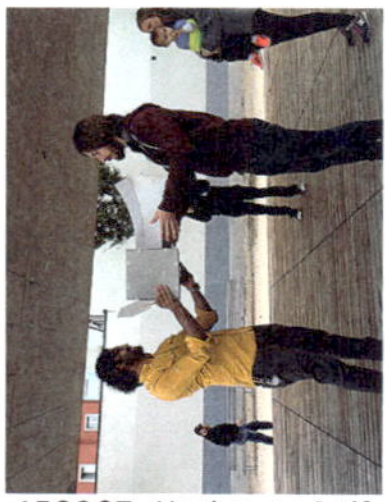
150907_Hothouse4.tif

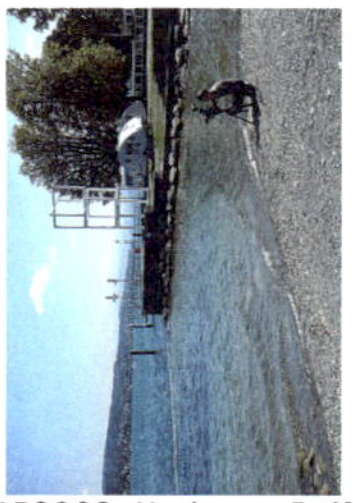
150908_Hothouse5.tif

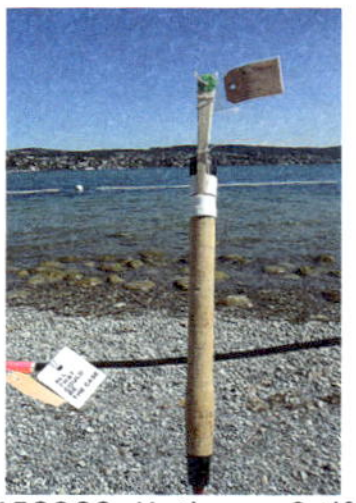
150908_Hothouse6.tif

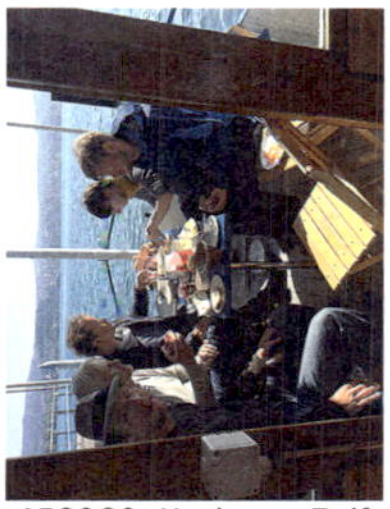
150908_Hothouse7.tif

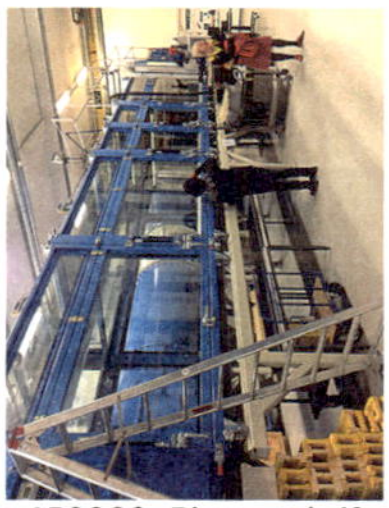
150908_Physmod.tif

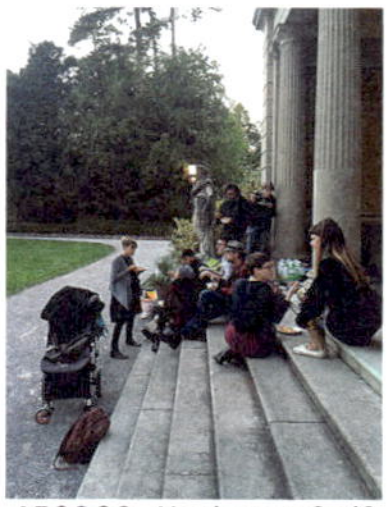
150909_Hothouse8.tif

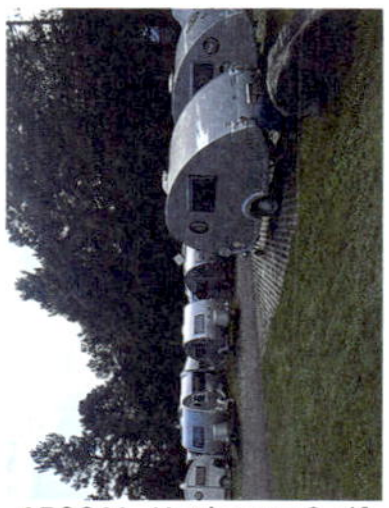
150911_Hothouse9.tif

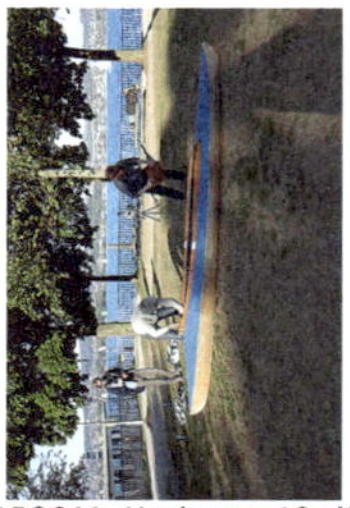
150911_Hothouse10.tif

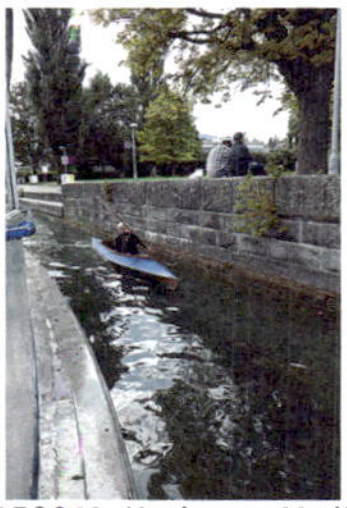
150911_Hothouse11.tif

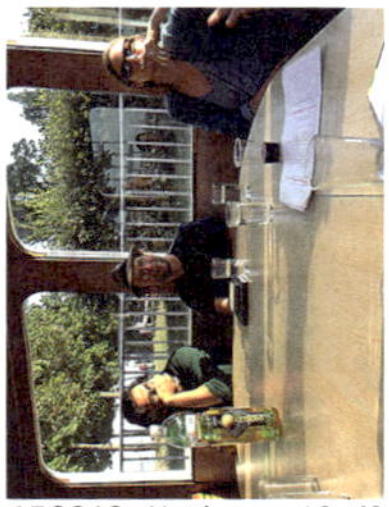
150912_Hothouse12.tif

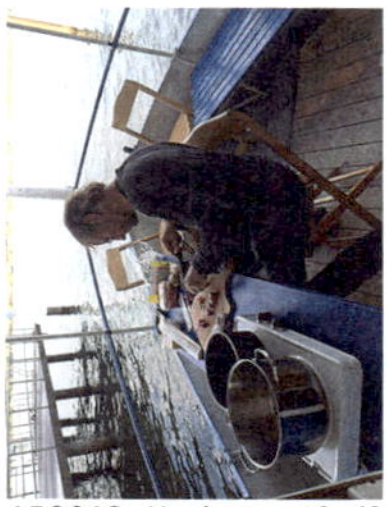
150912_Hothouse13.tif

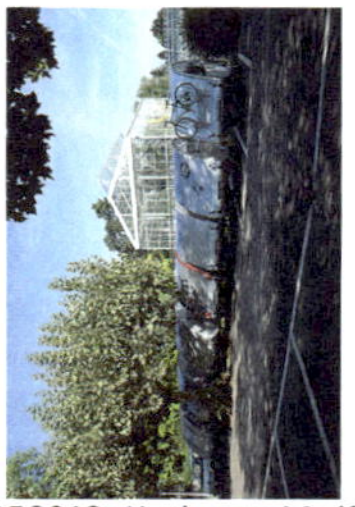
150912_Hothouse14.tif

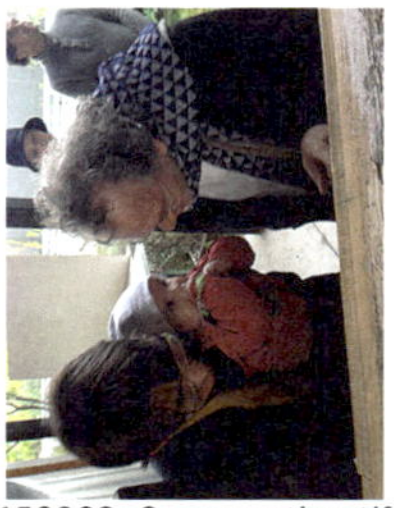
150922_Conversation.tif

Jul 2015
Listening

During the summer period the roof garden of the Toni Areal offered a pale green, fragrant and fairly quiet space. It attracted visitors during the day and far into the night. The wind tunnel then became a place for long talks that were occasionally interrupted by the night watch. In the early hours of the morning, when the plants were still wet with dew and only a few birds were singing, one might see a person who had spent the night in the tunnel sneak out, enjoy the view and listen to the city waking up. During the summer the work in the outdoor laboratory was more relaxed and KK set up a microphone, a small mixer and a camera. He listened to the wind and filmed the test section as a stage for the sound.

Kaspar König:
If you win wind, you lose silence.

Aug 2015
Instrument

A series of experiments in sound and wind led to the production of a short video on *The Sound of the Domesticated Wind*. The film was conceived by KK with Andrew Philips, Eirini Sourgiadaki and Yan Junjie. The 10'17" film uploaded to YouTube depicts the tunnel as an expanded musical instrument. The tunnel provided wind for the testing of the microphones. At which speed would the microphones clip? The sound of the microphones changed as they were lowered into tubes, in turn becoming instruments. Later the strings of a piano vibrated in the wind and the recorded tones reverberated through the tunnel.

150819
Conversation
Lena Unger, carpenter and business activist, has lunch at the wind tunnel and tests its whirling space.

Lena Unger:
To me the wind tunnel is an instrument which awakens the joy of sensuous experimentation.

Sep 2015
Hothouse

The second Hothouse (for Hothouse 1, see October 2013) was performed on the theme of scaling. For a week the team used the area along and above the western shore of Lake Zurich as their workspace, sliding between being a stranger and feeling at home. Each team member slept in a streamlined retro camper that was regularly moved to new settings. They washed themselves in the lake or used public showers. Meetings were held in Wollishofen, on the ship *Stäfa*. The team enjoyed their meals on the ship, but cleaned the dishes and fetched the water in the community centre 50 metres away. Throughout the week there was an exchange of assignments on the theme of scale. As with the Hothouse 2013, Piet Esch produced a film. Conducting research in the open air was significantly different and surely had an impact. However, this 'upscaling' demanded more energy and more planning. It simultaneously distracted and focused the team.

150906–150913
Hothouse
The second Hothouse is dedicated to scale (cp. October 2013). Guests include: Jochen Gerz, artist, Ireland; Alex Chitty and Daniel G. Baird, artists, Chicago; and Aleksander Gabrys, composer, contrabassist and performer, Basel.

150907
Conversation
Manuel Scotoni, student at ZHdK, passes by and engages in a talk about the tunnel.

150908
Publication
The group introduces the ZHdK wind tunnel at the *PHYSMOD 2015: The International Workshop on Physical Modelling of Flow and Dispersion Phenomena*, hosted by the Swiss Federal Institute of Technology Zurich/Empa Urban Physics Group.

150915
Conversation
The Reflection Group meets for the third time and discusses different ways of defining the project.

150922
Conversation
Georges Didi-Huberman, art historian and professor from Paris, gives a lecture at ZHdK and visits the wind tunnel.

Manuel Scotoni:
You feel different standing in the wind.

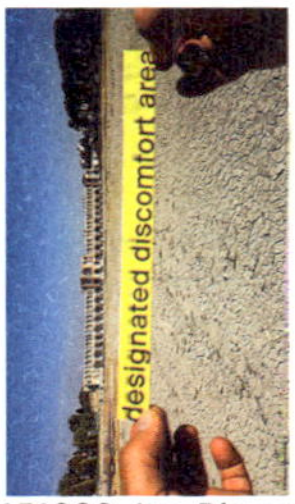

***_151000_IranDiscomf.tif

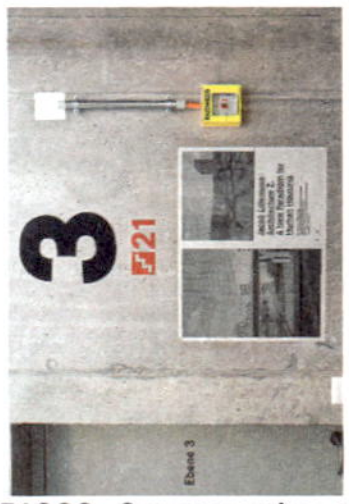
151022_Conversation.tif

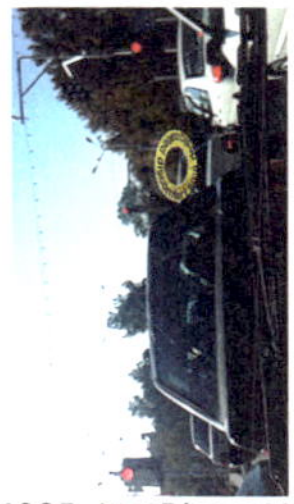
151025_IranDiscom1.tif

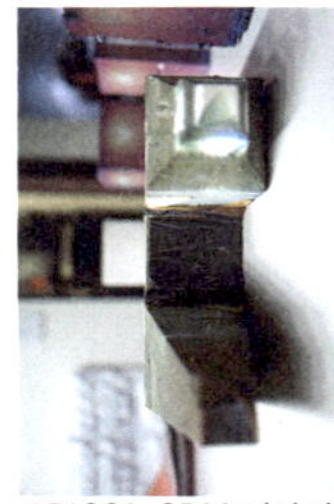
151031_3DModel.tif

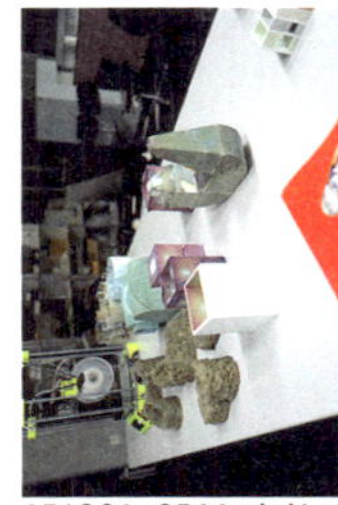
151031_3DModel1.tif

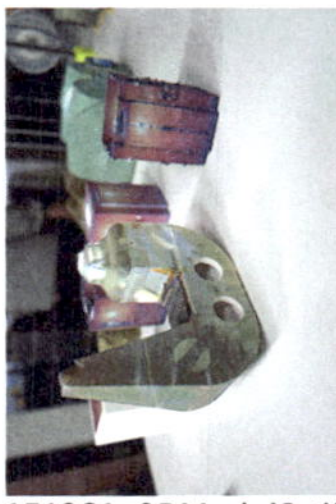
151031_3DModel2.tif

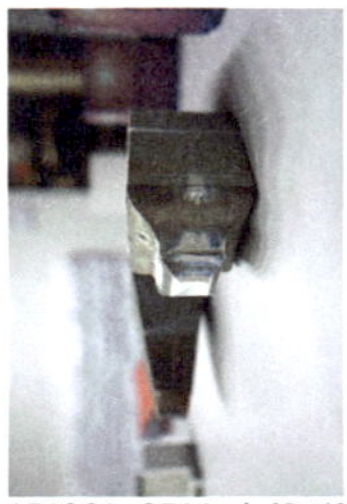
151031_3DModel3.tif

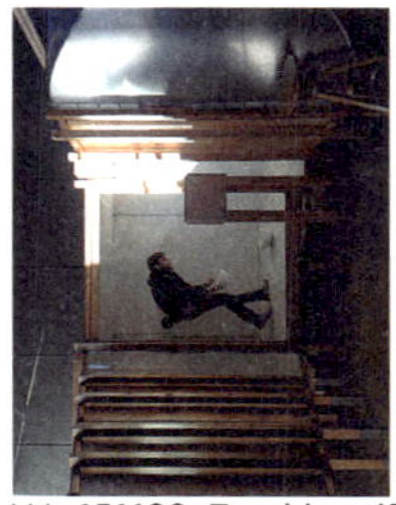
***_151100_Teaching.tif

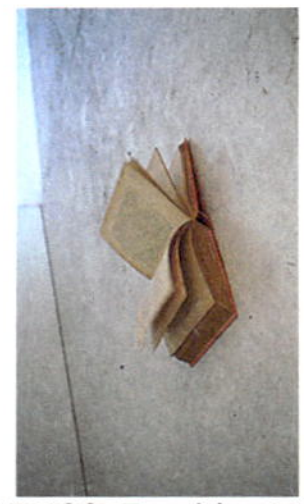
151106_Teaching1.tif

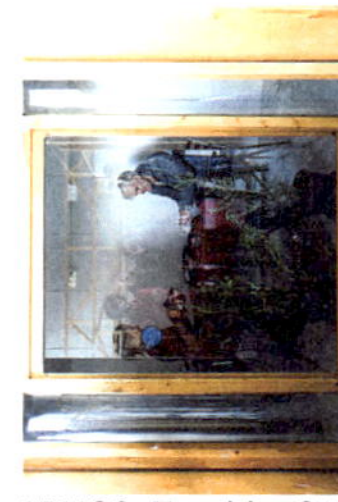
151106_Teaching2.tif

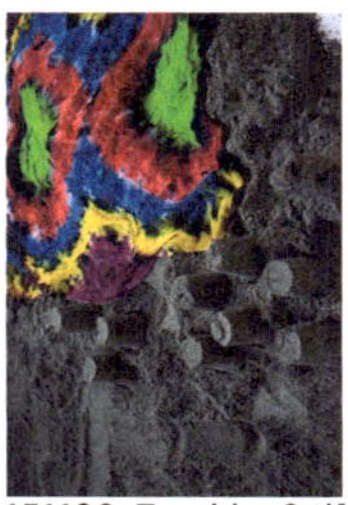
151106_Teaching3.tif

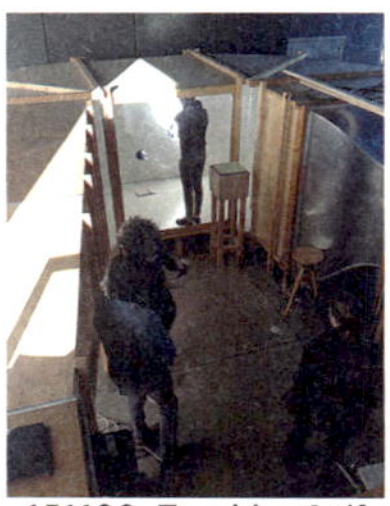
151106_Teaching4.tif

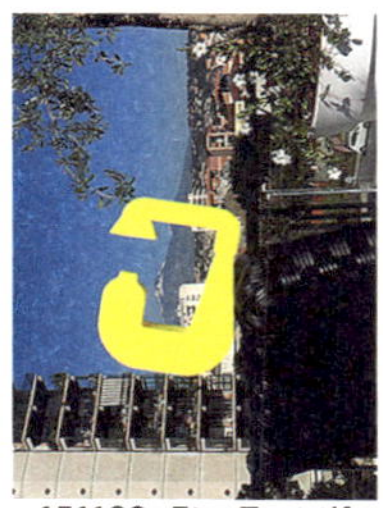
151109_EtnaTest.tif

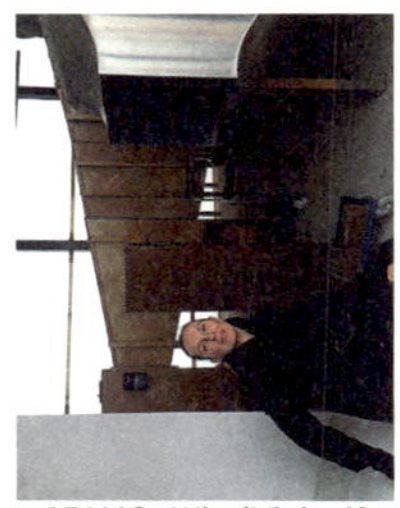
151112_WindVisit.tif

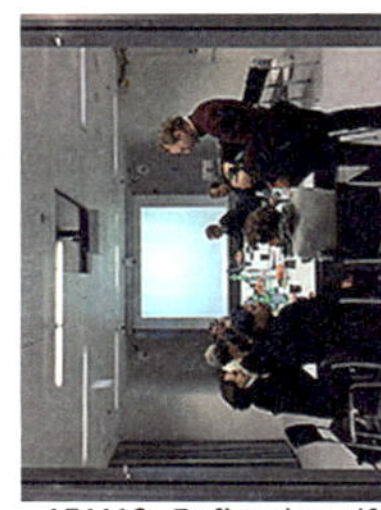
151118_Reflection.tif

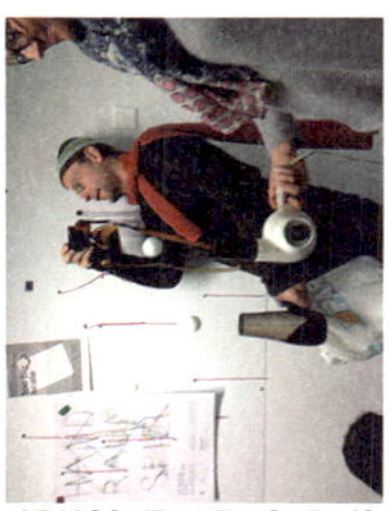
151123_TestForSAR.tif

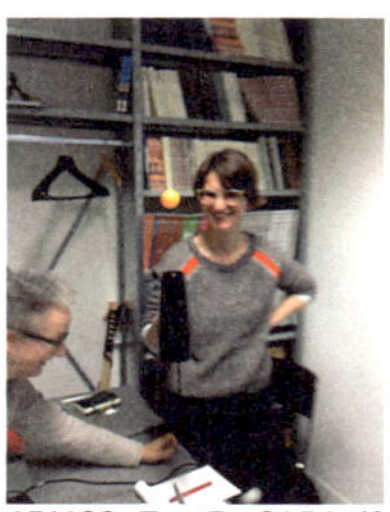
151123_TestForSAR1.tif

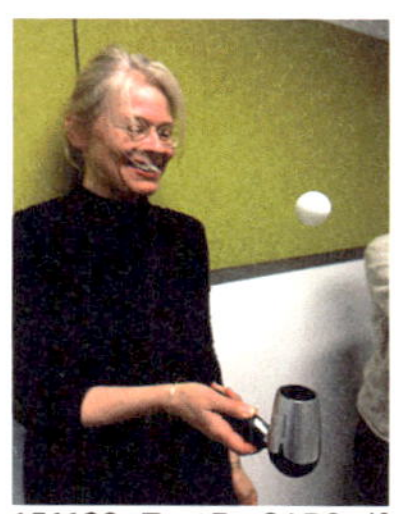
151123_TestForSAR2.tif

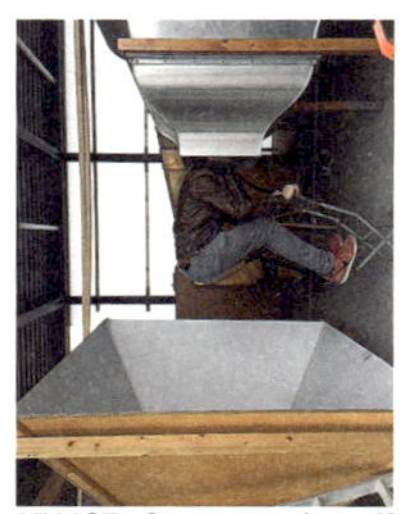
151127_Conversation.tif

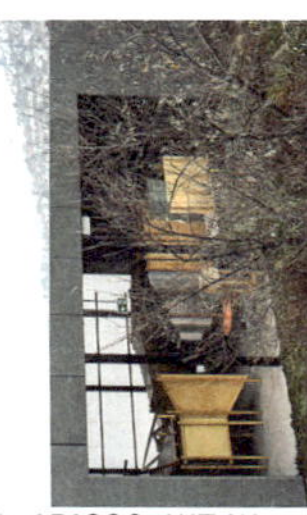
***_151200_WT-Wood.tif

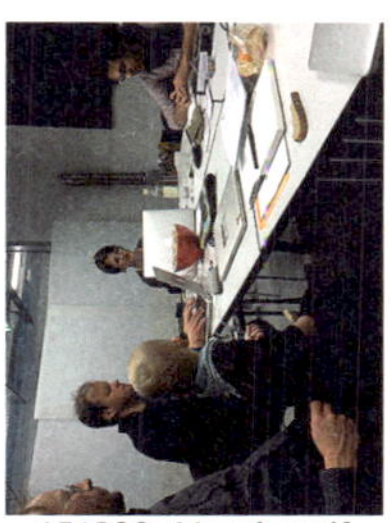
151209_Meeting.tif

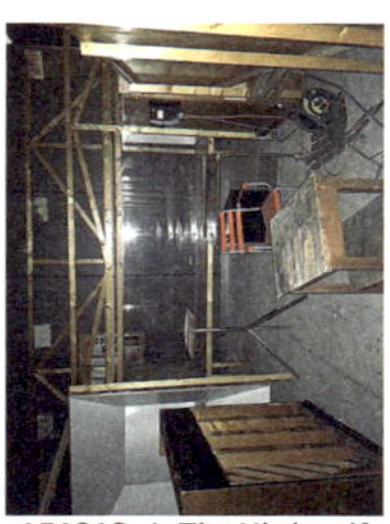
151212_InTheNight.tif

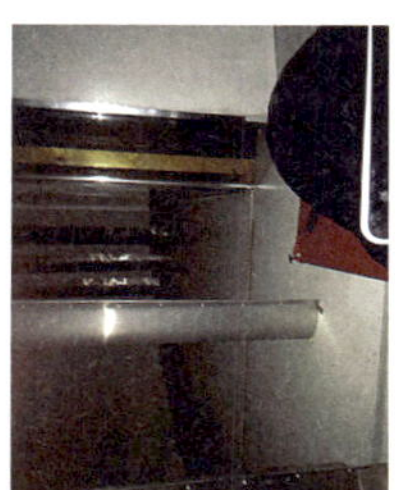
151212_InTheNight1.tif

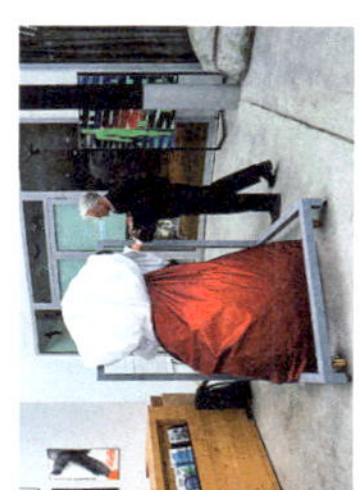
151212_WT-Air.tif

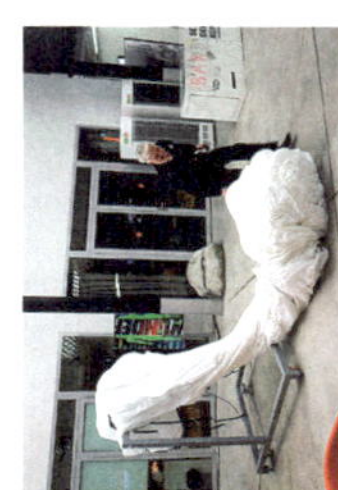
151212_WT-Air1.tif

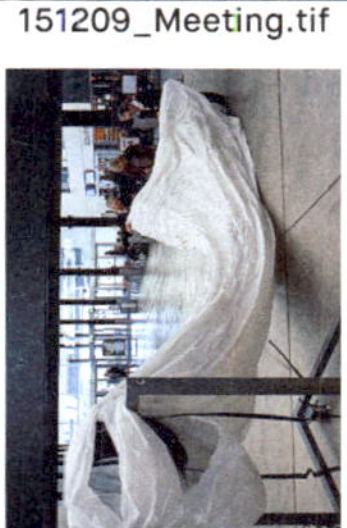
151212_WT-Air2.tif

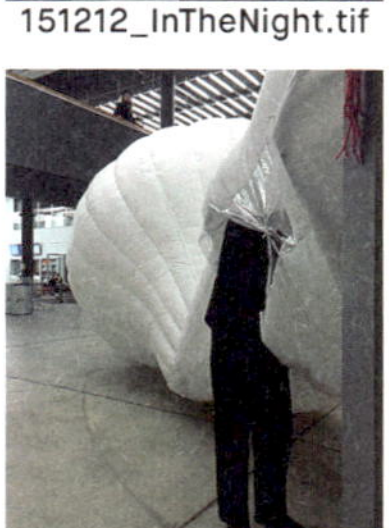
151212_WT-Air3.tif

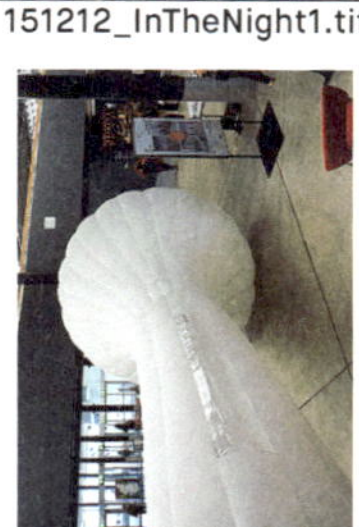
151212_WT-Air4.tif

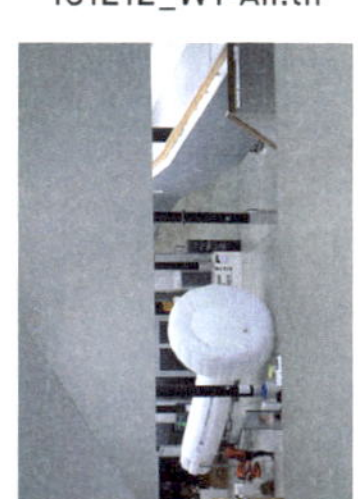
151212_WT-Air5.tif

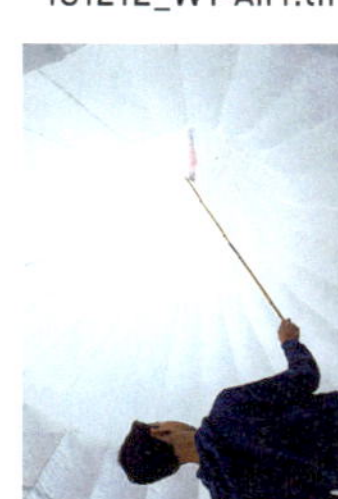
151212_WT-Air6.tif

Oct 2015
Resonance

Like Sahara dust or birds of passage the materials of the wind tunnel project, 3D miniature models and its other multiples travelled around the globe and turned up in places where one would not immediately expect them, thus transferring the project's physical traces on to various locations. From Gothenburg to Catania and Malta, objects were included in talks and presentations and provoked audiences' desire to have a similar wind tunnel as a tool for negotiating issues on their own agendas.

151012
Meeting
All team members write a page on the role of the wind tunnel in their work. This input also functions as a prototype for the elevator pitches in this book.

151022–151023
Conversation
Jacob Lillemose, critic and curator from Copenhagen, visits the wind tunnel and gives a lecture about Zombie Safe Houses.

Nov 2015
Studies

The space of the wind tunnel gradually became a place to meet and work. A place you could move into and reflect upon your studies or work alongside or inside. The students of the MA Transdisciplinary Studies regularly collaborated around and used the tunnel as part of their work. The project's PhD students, SW and HA, taught courses in Basel and Zurich corresponding to their particular lines of inquiry integrating specific experiences with the tunnel. For instance, SW focused on the academic analysis of models in film and HA on the artistic concept of the 'Wind Tunnel World'.

151102–151106
Teaching
HA starts a course on remote sensing in the MA Transdisciplinary Studies programme and presents his PhD research to students.

151112
Meeting
Conceptual work on the *Wind Tunnel Bulletin #5*, to be published in the summer of 2016.

151118
Conversation
The final meeting of the Reflection Group focuses on the formats of the project's publications.

151127
Conversation
Ronny Hardliz – artist, PhD candidate and SNSF Doc.Mobility-fellow – visits the wind tunnel.

Dec 2015
Unlikely Twins

At the ZHdK's Research Day the team presented the wind tunnel's unlikely twin. The *Inflatable Wind Tunnel* (2015) is an artwork by KK and FD that builds up its outer structure, inflating it with the leftover air from the inside of its wind-tunnel test section. Conceptually it follows the Göttingen-type wind tunnel, which returns the air from the blowing to the sucking section – like the circular model which the wind tunnel developed into during the project. The new inflatable wind tunnel, however, has the straight shape of an Eiffel type tunnel, as did the first version of the Hafnerstrasse tunnel (see illustration September 2013). With its integrated air circulation, the inflatable becomes its own means of support. It also has a closed testing area. Who would have thought that, less than three years after the first wind tunnel was built in the Hafnerstrasse, an artwork with such a clear and soft frame would unfold within a research project that had produced so much disturbance and smoke?

151209
Meeting
Working on the book of *the too big to scale* symposium (see May 2015), which will be published in 2017.

151212
Publication
At the annual Research Day of the ZHdK the *Inflatable Wind Tunnel* by KK and FD is presented to the public for the first time.

151215
Meeting
To conclude the project, the team meets for dinner on one of Basel's ferries, travelling back and forth across the Rhine.

151219
Multiple
The team's Christmas presents to the ZHdK administration are gingerbread cakes decorated with the silhouette of the wind tunnel and an incense stick, promising sweetness, smoke and fire.

151231
The official end of the SNSF project 'Size Matters: On Scale and Size of Models'.

The following publication list is a selection of output of the members of Research Focus in Transdisciplinarity since the start of the research project 'Size Matters. On Scale and Size of Models' in January 2013. The list makes no difference between scientific and artistic output. The selected publications are all related to the wind tunnel activities and are sorted according to their main topic. Nevertheless many of the publications could have been listed in several categories. Dates are indicated in the yyyy, dd-mm format.

1. On Models

1.1 Models and Theory

Dombois, Florian, 2013, 'Models to Think' in Kut, Elvan, and Martin Schmid (eds.), *Heilen – Gesunden: Das andere Arzneibuch. Festschrift zum 60. Geburtstag von Gerd Folkers*. Zurich: Edition Collegium Helveticum: 103–108. (Picture essay)

Dombois, Florian, 2015, *Language of Things*. (Art work, multiple)

Dombois, Florian, 2015, 09.11.–11.11., 'Language of Things' at *Proto-Objects*, De Bijloke, Gent (BE). (Group exhibition)

Wendler, Reinhard, 2013, *Das Modell zwischen Kunst und Wissenschaft*. München: Wilhelm Fink. (Monograph)

Wendler, Reinhard, 2013, 14.02., 'Rivals, Catalysts, Accomplices' at *Conference of the College Art Association*, New York (USA). (Conference, lecture)

Wendler, Reinhard, 2013, 16.09., 'Scale Model Future Knowledge' at Symposium *Matters of thought: models and knowledge-production in the early-modern world*. Royal Society, London (GB). (Conference, lecture)

Wendler, Reinhard, 2014, 'Es gibt Dinge, die dulden keine Herstellung in einem Modell. Zur Rolle der Geltung von Entwurfsmedien' in *Archiv für Mediengeschichte*: 73–83. (Essay)

1.2 Models and Film

Burr, Martin, 2013, 25.08., 'Film' at *Conference International Cartographic Association/Association Cartographique Internationale*, Dresden (D). (Workshop)

Waltenspül, Sarine, 2013, *Der Reiz der Uneindeutigkeit. Modelle im Film zwischen Technik, Ästhetik und Illusion*. Zurich: Zurich University. (Master thesis)

Waltenspül, Sarine, 2014, 'Medienreflexion im Kippmoment: Kinematografische Modelle' in *Das Wissen der Künste* 3, Berlin (Online). (Essay)

Waltenspül, Sarine, 2016, 29.11., 'Kinematografische Modelle: Darstellungstechniken im Transfer' at *Jahrestagung Gesellschaft für Modellforschung*, Humboldt-Universität, Berlin (D). (Conference, lecture)

1.3 Models and Remote sensing

Ahmed, Haseeb, 2013, 23.05.–30.05., 'Miniature World' at *Lahore Literary Arts Festival*, Alhamra Arts Center, Lahore (PAK). (Art work, installation)

Ahmed, Haseeb, 2015, 05.08., 'Artist Talk' at *7th Annual Alanica Symposium*, North Ossetia (RUS). (Performance)

Ahmed, Haseeb, 2015, 07.10., 'Artists Talk' at John Lennon School of Art and Design, Liverpool (UK). (Performance)

Dombois, Florian, 2014, 27.03., 'Artist Talk' at John-Hopkins-University, Baltimore (USA). (Performance)

1.4 Models and Architecture

Ahmed, Haseeb, 2014, 'On the Relationship of Form and Identity in Midwestern Islamic Architecture' in *The New Literary Observer (Moscow)*: 299–312. (Essay)

Ahmed, Haseeb, 2015, 'Descending Through Densities' in ExtraCity Kunsthal Antwerpen (ed.), *Allegory of the Cave Painting Reader*. Mailand: Mousse Publishing: 291–309. (Essay)

Allen, Jane Ingram, 2014, 'Review of Florian Dombois *uboc No.1 & stuVi2* at Boston University' in *Sculpture Magazine* 4: 73. (Press review)

Cardon, Stephanie, 2013, 'uboc No.1 & stuVi2: A Conversation with Florian Dombois' in *Big, Red & Shiny* 2 (14) (Online). (Press review)

David & Goliath. Modelle zwischen Kunst und Architektur, 2013, 22.11., organised by Florian Dombois, Stefan Trüby and Reinhard Wendler at Zurich University of the Arts, Zurich (CH). (Conference, lectures and performances)

Dombois, Florian, 2014, *Zugabe* at Courtyard of Brandenburg Parliament, Stadtschloss Potsdam (D). (Art work, permanent installation)

Dombois, Florian, 2013, *uboc No.1 & stuVi2* at Campus of Boston University, Boston (USA). (Art work, temporary installation)

Dombois, Florian, 2013, 24.04., 'Models' at ACT, Massachusetts Institute of Technology (MIT), Cambridge (USA). (Performance)

Dombois, Florian, and Luis Berríos-Negrón, 2015, 18.05, 'Remote-Sensing, Landscape, Scale' at *Rethinking Research Practices*. Royal Institute or Technology (KTH) and Konstfack – University College of Arts, Crafts and Design, Stockholm (S). (Conference, performance)

Müller, Josef Felix (ed.), 2014, *Florian Dombois: Zugabe*. St. Gallen/Berlin: Vexer-Verlag. (Monograph)

Wendler, Reinhard, 2014, '1:1 oder der Abgrund des Doppelpunkts/1:1 or: The Chasm of The Colon' in Lange, Christiane, and Robbrecht et Daem architecten (eds.), *Mies 1:1. Das Golfclub-Projekt*. Cologne: Buchhandlung Walter König: 126–135. (Essay)

1.5
Models and
Biology

Ahmed, Haseeb, 2013, 'Fish Bone Chapel'. (Art work, installation)

Ahmed, Haseeb, 2013, 14.09.–31.03., 'Fish Bone Chapel'at Naturalis – Natural History Museum, Leiden (NL). (Solo exhibition)

Ahmed, Haseeb, 2015, 29.03.–29.04., 'False Alternatives' at Veda Gallery, Mumbai (IN). (Art work, installation)

Wendler, Reinhard, 2013, 'Visuelles Denken in James Watsons *Molecular Biology of the Gene*' in Mareis, Claudia, and Christof Windgätter (eds.), *Wechselbeziehungen zwischen Design-, Medien- und Wissenschaftsforschung*. Berlin: Diaphanes: 23–37. (Essay)

1.6
Models and
Geology

Dombois, Florian, 2013, 14.05., 'Landforms' at Sint Lucas School of Art and Design, Antwerpen (BE). (Keynote lecture)

Wendler, Reinhard, 2013, 'Epistemische Ballistik und die Rhetorik der Unverhältnismässigkeit' in Dombois, Florian, Priska Gisler, Schirin Kretschmann and Markus Schwander (eds.), *Präparat Bergsturz II*. Luzern: Edizioni Periferia: 57–67. (Essay)

2. On Scale

2.1
Scale and
Theory

Ahmed, Haseeb, Martin Burr, Florian Dombois, Julie Harboe, Kaspar König, Mirjam Steiner and Sarine Waltenspül, 2015, 15.–18.06., 'Size Matters. On Size and Scale of Models' at *Scale. 2015 Conference of the European Society for Literature, Science, and the Arts*, Malta (M). (Conference lecture)

Dombois, Florian, and Julie Harboe (eds.), 2017, *Too Big To Scale. On scaling space, numbers, time and energy.* Zurich: Scheidegger & Spiess. (Monograph)

too big to scale. Ein interdisziplinäres Symposium zu Fragen der Skalierung, 2015, 08.–09.05., organized and hosted by Research Focus in Transdisciplinarity at Zurich University of the Arts, Zurich (CH). (International conference)

Wendler, Reinhard, 2017, 'Orders of Magnitude in Scaling Effects' in Dombois, Florian, and Julie Harboe (eds.), *Too Big To Scale. On scaling space, numbers, time and energy.* Zurich: Scheidegger & Spiess: 65–80. (Essay)

Wind Tunnel Bulletin #2. The Scale Effect. 2014 (June). Zurich. (Journal issue)

2.2
Scale and
Film

Waltenspül, Sarine, 2015, 'Der Kosmos aus der Petrischale. Oder: Wie lässt das Ähnlichkeitsprinzip Universen entstehen?' in Jürgens, Anna-Sophie, and Tassilo Tesche, *LaborARTorium*. Bielefeld: Transcript Verlag: 109–121. (Essay)

Waltenspül, Sarine, 2017, 'The Camera as a Scaling Instrument: Focus on Cinematographic Modelling Techniques' in Dombois, Florian, and Julie Harboe (eds.), *Too Big To Scale. On scaling space, numbers, time and energy.* Zurich: Scheidegger & Spiess: 33–48. (Essay)

2.3
Scale and
Art

Sandhana, Lakshmi, 2014, 'Built to Scale' in *Wired Magazine*: 74. (Press review on Haseeb Ahmed)

Steiner, Mirjam, 2017, 'Inseparably United: Big and Small in Literary Settings' in Dombois, Florian, and Julie Harboe (eds.), *Too Big To Scale. On scaling space, numbers, time and energy.* Zurich: Scheidegger & Spiess: 97–112. (Essay)

2.4 Scale and Time

Dombois, Florian, 2014, *Spaces Out Of Delay*. (Art work, installation)

Dombois, Florian, 2014, 19.09.–07.12., 'Spaces Out Of Delay' at *Allegory of the Cave Painting*, Extra City, Kunsthal Antwerpen (BE). (Group exhibition)

Dombois, Florian, 2015, 30.05., *Inverse – A Flock of Happenings* with contributions of Helmholtz-Zentrum Dresden-Rossendorf, MPI for Physics of Complex Systems, Center for Regenerative Therapies Dresden, MPI of Molecular Cell Biology and Genetics, TU-Dresden and Hochschule für Bildende Künste at Postplatz, Dresden (D). (Art work, happening)

Dombois, Florian, 2016, 30.10., *Inverse* at Zwinger Dresden (D). (Art work, happening)

König, Kaspar, 2013, 31.08., 'Slowbiking Contest Luzern' at *Industriestrassenfest*, Lucerne (CH). (Performance)

König, Kaspar, 2015, 21.–22.03., 'Slowbiking Contest Berlin' at *Veloberlin*, Berlin (D). (Performance)

3. On Wind Tunnels

3.1 Wind Tunnels and Transdisciplinarity

Ahmed, Haseeb, Martin Burr, Florian Dombois, Julie Harboe, Kaspar König, Mirjam Steiner and Sarine Waltenspül, 2015, 08.09., 'Wind Tunnel Bulletin Zurich' at *PHYSMOD – International Workshop for Physical Modelling of Atmospheric Flow and Dispersion in Fluid Modelling Facilities, such as Wind Tunnels and Water Flumes*, EMPA Zurich (CH). (Conference lecture)

Dombois, Florian, and Sarine Waltenspül, 2013, *The Endless (Wind) Tunnel*. (Art work, video)

Dombois, Florian, and Sarine Waltenspül, 2016, 15.07.–08.01.2017, 'The Endless (Wind) Tunnel' at *Welten bauen. Modelle zum Entwerfen, Sammeln und Nachdenken*, Museum für Gestaltung, Zurich (CH). (Group exhibition)

Dombois, Florian, 2014, *Tell Tales*. (Art work, multiple)

Modellapéro. Ein Open Studio der Forschung mit Windkanal, Buchvernissage und Musik, 2013, 21.11., at Zurich University of the Arts, Zurich (CH). (Installation)

Modellapéro. Ein Open Studio der Forschung mit Windkanal und Musik, 2014, 26.06., at Zurich University of the Arts, Zurich (CH). (Installation)

Modellapéro. Ein Open Studio der Forschung mit Windkanal, Musik und Buchvernissage, 2015, 25.06., at Zurich University of the Arts, Zurich (CH). (Installation)

Modellapéro. Ein Open Studio der Forschung mit Windkanal, Musik und Buchvernissage, 2016, 15.06., at Zurich University of the Arts, Zurich (CH). (Installation)

Wind Tunnel Bulletin #1. 2013 (November). Zurich. (Journal issue)

Wind Tunnel Bulletin #3. 2014 (December). Zurich. (Journal issue)

Wind Tunnel Bulletin #5. 2016 (June). Zurich. (Journal issue)

Wind Tunnel, 2014, 27.09.–28.09., at 'ZHdK Tage der offenen Tür', Zurich University of the Arts, Zurich (CH). (Installation)

Wind Tunnel Smoke, 2014, 25.10., at 'ZHdK Party – Creative City', Zurich University of the Arts, Zurich (CH). (Installation)

3.2 Wind Tunnels and Art

Ahmed, Haseeb, 2014, 04.12.–04.01.2015, *A is for Albatross* at Museum Bärengasse, Zurich (CH). (Solo exhibition)

Ahmed, Haseeb, 2016, 04.05., *Aviary* at Von Karman Institute for Fluid Dynamics, Rhode-St-Genèse (BE). (Solo exhibition)

Ahmed, Haseeb, 2016, 08.09.–08.10., *Wird* at Harlan Levey Projects, Brussels (BE). (Solo exhibition)

Ahmed, Haseeb, 2016, 16.09–18.09., *Wind Works* at Counter Space, Zurich (CH). (Solo exhibiton)

Dombois, Florian, 2013, *Wind Tunnel Drawings*. (Art work, drawings)

Dombois, Florian, 2014, 19.10.–14.12., 'Wind Tunnel Drawings' at *Round the Corner*, Maison Grégoire, Brüssel (BE). (Group exhibition)

Dombois, Florian, 2016, 03.07.–28.08., *Inside Out* at Kunsthalle Marcel Duchamp | The Forestay Museum of Art, Cully (CH). (Solo exhibition)

König, Kaspar, and Florian Dombois, 2015, *Inflatable Wind Tunnel*. (Art work, sculpture)

König, Kaspar, and Florian Dombois, 2016, 15.07.–08.01.2017, 'Inflatable Wind Tunnel' at *Welten bauen. Modelle zum Entwerfen, Sammeln und Nachdenken*, Museum für Gestaltung, Zurich (CH). (Group exhibition)

Steiner, Mirjam, 2016, 'How Worlds Come to Existence. Nostalgic and Virtual Model Worlds at the Design Museum' in *Zett* 1, 56–59. (Press review)

3.3 Wind Tunnels and Politics

Ahmed, Haseeb, 2013, 07.12., 'Remote Sensor: EU Parliament Wind Nuissance Study' at *LaborARTorium*, Ludwig Maximillians-Universität, Munich (D). (Conference, performance)

Ahmed, Haseeb, 2017, 'Wind Tunnel World: Where the Wind Blows Straight' in Dombois, Florian, and Julie Harboe (eds.), *Too Big To Scale. On scaling space, numbers, time and energy.* Zurich: Scheidegger & Spiess: 177–192. (Essay)

4. On Artistic and Scientific Modes of Working

Dombois, Florian, 2013, *Venus und Mathemata – Eine Doppelprojektion*. (Art work, installation)

Dombois, Florian, 2013, 06.06., *Venus und Mathemata – Eine Doppelprojektion* at Center for Interdisciplinary Research, Bielefeld (D). (Solo exhibition)

Dombois, Florian, 2016, 26.11., 'Art with some T' at *Not only between, but even beyond. Oder: Transdisziplinarität – eine Bestandsaufnahme* at Collegium Helveticum, Zurich (CH). (Conference, performance)

Esch, Piet, with Florian Dombois, 2014, 18.12., *Hothouse Uerikon (2013)*. Premiere at Cinema Xenix, Zurich (CH). (Documentary film)

Esch, Piet, with Florian Dombois, 2017, 27.04, *Hothouse #2 (2015)*. Premiere at Cinema Rote Fabrik, Zurich (CH). (Documentary film)

Harboe, Julie, 2015, 'Werk und Prozess' in Badura, Jens, et al. (eds.), *Künstlerische Forschung. Ein Handbuch*. Berlin: Diaphanes: 243–247. (Essay)

Harboe, Julie, 2015, 30.04.–01.05., *Unconditional Love. Spring Event of Society for Artistic Research*. Chelsea College of the Arts and University of the Arts, London (GB). (Conference, panel chair)

Harboe, Julie, 2015, 21.–22.05., 'Worlding: Scaling Collaborations of Artistic Research and Transdisciplinary Methodologies' at *transvaluation*, Chalmers University of Technology, Göteborg (S). (Conference paper)

Harboe, Julie, 2016, 04.–07.05., *3rd World Ornamental Forum*. Collaboration Future Laboratory CreaLab, LUASA and Kirchner Museum Davos (CH). (Conference co-organization, video essay)

Harboe, Julie, 2017, 'On Human Scale and Calibrating Research Groups in a Wind Tunnel' in Dombois, Florian, and Julie Harboe (eds.), *Too Big To Scale. On scaling space, numbers, time and energy.* Zurich: Scheidegger & Spiess: 113–128. (Essay)

König, Kaspar, 2014, 21.02., *Versuchsanordnung 3 / Transform. Ein interdisziplinäres Kunstprojekt über 7 Wochen*, Güterstrasse 8, Bern (CH). (Concert)

Wind Tunnel Bulletin #4. Fahrkunst. 2015 (June). Zurich. (Journal issue)

5. On Aesthetic Thinking

5.1. Aesthetic Thinking and Theory

Dombois, Florian, Mira Fliescher, Dieter Mersch and Julia Rintz (eds.), 2014, *Ästhetisches Denken. Nicht-Propositionalität, Episteme, Kunst.* Zurich: Diaphanes. (Monograph)

Dombois, Florian, 2013, 'Luginsland (On Art as Research)' in Bauer, Ute Meta, and Thomas D. Trummer (eds.), *AR – Artistic Research*. London: Koenig Books: 76–85. (Essay)

Dombois, Florian, 2013, 'Kunst als Forschung. Ein Versuch, sich selbst eine Anleitung zu schreiben' in Hedinger, Johannes M., and Torsten Meyer (eds.), *What's next? Kunst nach der Krise*. Berlin: Kadmos: 139–143. (Essay, reprint from 2006)

Nichtpropositionalität und ästhetisches Denken, 2013, 21.05., organized and hosted by Research Focus in Transdisciplinarity at Zurich University of the Arts, Zurich (CH). (International conference)

5.2 Aesthetic Thinking and Teaching in the Arts

Dombois, Florian, 2013, 29.03., 'The Ill-Mannered Daughter' at *SeminArt#9* at Higher Institute for Fine Arts Ghent (BE). (Keynote lecture)

Dombois, Florian, 2014, 14.04, *Peter holds Dorothy* at Academy of Fine Arts, Helsinki (FIN). (Lecture)

Dombois, Florian, 2014, 18.04., *Peter holds Dorothy. Why speaking, reading and writing as an artist* at Ecole Cantonale d'Arts, Sierre (CH). (Lecture and workshop)

Dombois, Florian, and Isabel Mundry, 2015, 27.05., *Was ist verhandelbar? Inhalte des Kompositionsunterrichts*, at Hochschule für Musik und Theater München (D). (Lecture)

5.3 Aesthetic Thinking and Online Publication

Mind the Gap! 2013, 07.–08.03., organised by Florian Dombois and Gerhard Eckel, Ligeti-Hall, MUMUTH, Graz (A). (International conference and concert series)

Harboe, Julie, 2015, 30.04., 'The Artistic Research Library' at *Unconditional Love. Spring Event of the Society of Artistic Research*, Chelsea College of the Arts, London (GB). (Conference, panel chair)

5.4 Aesthetic Thinking and Sound

Dombois, Florian, 2013, 04.04., *Angeschlagene Moderne* at Collegium Helveticum, Zurich (CH). (Keynote lecture)

Dombois, Florian, 2013, 11.12., *Surf* at Chair for Computer Aided Architectural Design, Department for Architecture, ETH Zurich (CH). (Concert and lecture)

Dombois, Florian, 2014, 05.06.–07.09., *Angeschlagene Moderne | Struck Modernism* at Museum Haus Konstruktiv, Zurich (CH). (Solo exhibition)

Dombois, Florian, 2015, 31.05., *What Are the Places of Danger?* at Research Pavilion, Venice (I). (Lecture)

Haldemann, Anita, 2014, 'Florian Dombois – Angeschlagene Moderne' in *Kunst-Bulletin* 9: 88–89. (Press review)

König, Kaspar, 2013, 28.06., at *next_generation 5.0 AKUSMATIK* at ZKM Karlsruhe (D). (Concert)

König, Kaspar, 2014, 14.02.–04.05., 'instant display. Klanginstallation für einen Flügel' at Stadtgalerie Neukölln, Berlin (D). (Group exhibition)

König, Kaspar, 2014, 13.06.–14.06., *Hackteria. Biologische Kunst, Wissenschaft, Natur und Biohacking*, at Museum zu Allerheiligen, Schaffhausen (CH). (Workshop section)

König, Kaspar, 2016, 16.12., *Beyond Sonification* at Kunstraum Walcheturm, Zurich (CH). (Sound festival)

Museum Haus Konstruktiv (ed.), 2014, Florian Dombois: *Angeschlagene Moderne | Struck Modernism* (exh. cat.). Berlin: The Green Box. (Monograph)

phi, 2014, 12.08., 'Wie klingt Kunst? Florian Dombois' Installation', *Neue Zürcher Zeitung*: 15. (Press review)

It's not about perpetual mobility; it's about a mode of qualifying actuality that continually discloses its procedures.

Vera Bühlmann, 2016

Not-yet art

When we publish something that is a success, it gives us satisfaction. Something seems to have been achieved; now something is said or shown the way it really is. Really at this moment and really in this place. For perhaps unlike some proponents of progress, I view publishing as a place- and time-dependent articulation, as a qualification of actuality. I know many natural scientists must be getting uneasy now: please, how can something stemming from the moment claim to be true? What will become of research if reproducibility is no longer at the centre? And I have to smile at this question. For I believe that in order to endure, something has to truly dwell in the time of its formulation and arise from it. And this is also how past artworks seem to impose themselves on us: it is precisely the ones that risked being ephemeral that attained intransience.

Since we can trace art all the way back to cave painting, while the sciences can perhaps look back on 'only' a five-hundred-year history, I put my faith in the older one. And of course I know that in the ephemeral, satisfaction is rare and pain stemming from the fact that what we had imagined to achieve has not yet been completely attained is much more common. That is the price for the bliss of effort. Opting for the ephemeral is a risky game, that is also clear. Most of it will be blown away and forgotten. But when you lapse into pettiness and give a pernickety account, you can be sure *a fortiori* that your own count will become obsolete sooner or later. So it's better to play, better to take risks.

Risking for example this idea: what do research results look like that an artist can use further? How can artist researchers organize and publicize their own material and methods such that independent artists as producers of art can adopt them in order to use them in their work? What kind of material would have higher potential for an artwork or for the development of art? And above all: what kind of not-yet art could possibly harbour art in itself as a potential? Indeed, shouldn't that be the real task of research at an art university: to publish things that give artistic production a boost? Shouldn't so-called artistic research first and foremost serve the arts themselves? In chemistry, for instance, there are research chemists and industrial chemists. The first do research, the second apply it. What would it mean to work in this way for the art market: that the researchers prepare and provide but do not yet transfer anything to the economic realm? A not-yet art. Or an art of the not-yet.

Proto-disciplinary publication formats

I'm sceptical about quality requirements from several disciplines simultaneously being placed on a publication. Only rarely can a good scholarly essay lay claim to being an artwork at the same time. But this not-yet art might be able to meet a not-yet science. Then, namely, when the disciplinary criteria have not yet taken hold, when the material is still mouldable and finds itself in a head-down embryonic position, as it were, then it seems to me to be possible to be attractive in several directions, for individual artists and individual scientists. And it is precisely this status that I repeatedly paid special attention to during the work on *The Wind Tunnel Model*. I would like to call these formats *proto-disciplinary*, for they offer themselves to different disciplines without clearly belonging to one discipline. Here are four examples of this that we believe it is worth developing further:

Example 1:
Dynamic picture panels

Even before the 'Size Matters' project began, the research focus team collected a thousand pictures from art and science on wind tunnels and models. Subsequently this number was reduced to 59 favourites, i.e. those pictures that all four of us deemed best. The process of selection and discussion lasted several hours and was documented on film.[1] The resulting selection was published online with nine different sorting suggestions.[2] The sorting suggestions also adhere to the 'not-yet' logic: the categories of the nine arrangements resulted from the material and intensive viewing of it without being clarified verbally. In the sorting process, the group tried nine times to perform a technique of picture arrangement inspired in Aby Warburg's *Mnemosyne Atlas* without agreeing on a conceptual category that would have been a criterion for this choice. Today if one selects one of the nine themes on the website, the pictures arrange themselves around a centre and get smaller and smaller towards the outside in five stages.[3] These arrangements are forms of condensation that impressed themselves forcefully on us out of the material after intense viewing.

Example 2:
Media archive

In a second attempt made around the same time with the same source material, we selected over a hundred pictures whose background information was essential for viewing the picture. We recorded the image and accompanying text in a media archive.[4] In the process, we experimented with various kinds of accompanying texts, ranging from explanatory descriptive approaches to poetically inducing ones. The German artist Harald Klingelhöller once called discourse the "plinth of the sculpture" (Reust 1997: 292, original in German) and, analogously, one of our aims was to reflect on the indispensable language work of the artist. What effect do accompanying texts have on the (poetic) potential of historical material or artworks? How do we build a discourse around the wind tunnel on which our and other artistic and scientific works can be built later? Might that perhaps be an essential task of artist research: to build up new discourses and to create new themes that will later impact the art market?

Example 3:
Modellapéro

Since 2012 we have organised an annual *Modellapéro*, a public event similar to an exhibition opening at which selected materials from our research are shown. On this occasion, we share parts of our work with the public on an equal footing. We do not explain, but rather we put materials and issues up for debate that we consider highly topical. Here, too, a balancing act is needed, between the condensation of the works as value creation and the simultaneous upholding of a stage of openness, the not-yet, to give others enough room for further processing of impressions and ideas, and further reflection. "No speeches are delivered", is therefore written on the invitation card each year.

Example 4:
Wind Tunnel Bulletin

Since 2013 we have published the semi-annual *Wind Tunnel Bulletin*, a kind of research letter. These bulletins are edited alternately by one half of our team and then the other half and are devoted to current material, questions and topics from our work. Both a manually produced print version and an online version appear.[5] The printed version is sold via international art book dealers, including *Printed Matter* in New York, *Motto Distribution* in Berlin and *Theophile's Papers* in Brussels and Paris. A few aerodynamic institutions also subscribe to it via the ZHdK. The *Bulletin* publishes texts and pictures in close cooperation with the Zurich-based designer Viola Zimmermann.

Artistic output

Apart from these attempts to open up for research a proto-disciplinary sphere that can deliver to both the arts and the sciences before being occupied by one of the disciplines, the members of the team also published their results in disciplinary formats. Their aim was to harness the potential of their joint work for their own respective discipline and to individually articulate what was developed collectively. For I do

1 Online at http://modulus.zhdk.ch/makingof.html (accessed 14 February 2017).
2 Online at http://modulus.zhdk.ch/ (accessed 14 February 2017).
3 We divided up according to: 1=belonging together, 2=related, 3=undecided, 4=quite far off, 5=not belonging.
4 We showed the collection printed on paper in a file during our first *Modellapéro* (29 November 2012) and in parallel put all the examples online at http://medienarchiv.zhdk.ch (see set 'Models with Stories', accessed 14 February 2017).
5 Online at http://windtunnelbulletin.zhdk.ch (accessed 14 February 2017).

not believe in a stable space in-between, in a sustainable language of transdisciplinarity per se. *Fahrkunst* is a transdisciplinary practice; it is rhythm and movement. But I don't think it is sensible to sediment it in a constant metamorphosis. Or let's just say, this is not our problem. I gladly leave this question to others.

The list above comprises a selection of what we have published in connection with our research at the wind tunnel to date. In compiling the list, I have suggested some possible comparisons that can be discussed. For in the context of national research funding we are increasingly subject to control based on evaluations of our success in a database. The counting methods of these databases follow the logic of (natural) sciences publishing. Now if as an artist I want to enter a disciplinary success of my work there, I have to adapt the categories to the specific criteria, values and qualities of my discipline. But – and this is of interest to me – the adaptations are not fanciful. For with a little openness scientific and artistic publishing can be compared astonishingly well when one transfers what was originally meant in the output category to the logic of artistic disciplines. In the following, I do this using Switzerland as an example. May my colleagues in other countries and with different funding categories examine and adapt this for their own respective situation.[6]

The expression of scientific research results in the categories of the Swiss National Science Foundation for output data is divided into:

(1) Scientific publications
(2) Academic events
(3) Knowledge transfer events
(4) Public communication
(5) Collaboration
(6) Use-inspired outputs
(7) Third-party funds
(8) Follow-up projects
(9) Awards

The first two categories are particularly important to the SNSF. With the scientific publications, a distinction is made between 'peer reviewed' and 'not peer reviewed'. And the second category, academic events, is subdivided into active 'participation' (talks given at a conference, posters, or individual talks) and 'self organized'. In other words, the ranking of the outputs in both formats is determined largely by the critical acclaim of peers. Mathematicians have to decide what good math is, and not a group of laypeople, to give an example.

And this is an interesting and important point: in the sciences the scientists themselves decide what will be researched; not politics, not business, not the public. Indeed, one could go so far as to say that research-funding institutions exist to protect the interests of scientific disciplines from the grasp of a short-term economy geared to return on investment. Basic research is needed, and now and then research aberrations. The actors of this research can decide what is worth investigating.

So our transfer of the concept of basic research to the arts seems to make sense: it is evident that it often takes a long time for ambitious art to unfold a broad effect. Van Gogh did not sell a single picture during his lifetime. Duchamp only really began his career, which today is inexorable, towards the end of his life; and it is anyone's guess who of today's artists will be considered a reference a hundred years from now. Thus we in the arts, just like in scientific research, are often confronted with long cycles of return on investment.

If we also want to transfer the concept of scientific peers to the arts, we have to be alert: peer review means, for instance, that a publication is evaluated by people who have the same expertise as those who wrote it. That would mean that an evaluation of artistic production could be best made by producers of art, i.e. artists. Today, this is common in only a few places of the art world. The art market, for example, is primarily determined by collectors and critics, who can appraise artworks but are not able to make art themselves. In some competitions, artists are called to sit on the jury, and some residency programmes are curated by artists. So if you want to set up a peer review for research at an art university, you have to push back the peers of reception and give sovereignty to the peers of production. This reminds me of so-called artists' artists, i.e. artists who are honoured by their colleagues long before the rest of the world recognises their value. Peer review aims precisely at such an evaluation: artists' artists and artists' art.

6 E.g. the Finnish Funding Agency TUKA mentions artistic output, but gives it it's own category, http://tutka.jyu.fi/help/english/publications/publicationtypes#F%20Public. See also http://tutka.jyu.fi/help/english/publications/definitions#artistic (accessed 14 February 2017).

Now what does that mean for research funding institutions? I don't think it is very productive to oblige the arts to produce knowledge following the patterns ruling the science circuit, as is very often argued in artistic research discourse today. I believe that generally speaking it is the task of funding institutions to enable simultaneous teaching and researching at universities. And this leads me to conclude that at art universities, which is where artists are trained, a kind of research has to be promoted that advances contemporary artists, 'fertilises' and accelerates their development. Therefore the judgement of the research projects – yet another conclusion – has to be made by peers of contemporary art, i.e. chiefly by other artists.[7] They have an explicit interest and the expertise to push forward artistic disciplines and thus also to strengthen art education.

I personally am open to interaction with the sciences. I consider an exchange and a mutual challenge to be extremely fruitful – otherwise I wouldn't have initiated projects such as the wind tunnel. But it is apparent that we cannot gauge the success of such cross-disciplinary projects based solely on scientific output. It is absolutely necessary that we also take into account artistic successes in parallel. And now I return to the initial question: if we take the nine categories the SNSF establishes regarding the sciences seriously, how could we categorise 'artistic output'? The following correspondences might be possible:

(1) Scientific publications = artistic publications
Monograph = exhibition of new works in a solo show / first performance of one or several compositions spanning a full-length concert evening / film presentation (premiere) / theatre performance (premiere) / monographic catalogue
Journal paper, contribution to an edited volume = exhibition of new works in a group show / first performance of a composition as part of a concert evening / film contribution (premiere) / theatre contribution (premiere)

I see as an artistic publication a single artwork shown or performed publicly, as well as an exhibition, a full-length concert, theatre performance or film event. We know that these formats are competitive, both in terms of the possibility to perform them and, later, of audience's judgement of them. Since we are active in the field of research, we are interested in evaluation by experts. One form of peer review is the invitation of an artist by another artist or a production oriented curator. I also consider a work that is selected as the winner of a competition by a competent jury to be peer reviewed. Commissioned works, depending on the degree of independence and importance of the commissioner, can also be categorized as peer reviewed. The reputation of the competition, the exhibition/concert/theatre/film institution or the curators is comparable with the ranking of scientific journals.

When an artist is given a solo show or a composer is given an entire evening for first performances of his compositions, this can be compared with a monograph. I would also put monographic catalogues in this category. When an artist is represented in a group exhibition, this is comparable to an essay in an edited volume. Premieres, first performances and first exhibitions count as original, repeat performances as reprints or events (see point 2). I would equate art-in-architecture artworks with patents, not only because larger amounts of money are involved, but also because the public is addressed for a longer time.

(2) Academic events = artistic events
Scientific talk = lecture performance / repeat performances of concerts, films, plays / exhibitions of works that have already been shown
Conference = festival
Poster = short exhibition / short performances

When visual artists hold a lecture-performance in front of their peer community, this can be likened to a scientific talk, for the work itself is presented with no mediation. When the lecture-performance takes place in the framework of a festival of peers, it is viewed as a talk given at a conference; but when it lasts a whole evening, it is considered an individual talk. I would also put repeat performances of concerts, films and plays in this category, as well as previews, or sketches of works still being articulated. Exhibitions installed for only one or a few

7 I would, however, by no means object to including such individual curators and other participants in contemporary art production who have demonstrably contributed to the production of works.

days can be compared to a scientific poster in my view.

(3) Knowledge transfer event = artist talk / podium / interview

When artists give a talk *about* their work, this is regarded as a knowledge transfer event, because the work is not actually presented, but discussed. The difference between a lecture and a lecture performance would be that the work is not performed during the lecture, but only described.

(4) Public communication = press reports

When an artist and his or her work are reported about in the media, this is considered public communication.

(5) Collaboration = collaboration

When cooperation e.g. with other artists results from the work, this is considered as collaboration.

(6) Use-inspired output = sale of an artwork

When an artwork arises in the context of a research project or is derived from one and is sold, this is comparable with use-inspired output.

(7) Third-party funds = new commission funds

When a research project results in paid artistic commissions, the amounts paid for the commissioned works are considered third-party funds.

(8) Follow-up projects = follow-up projects

When new artistic works or new research projects refer to a research project after the latter is finished, these are viewed as follow-up projects.

(9) Award = award

When an artist wins a competition or a prize, this is considered an award.

I do not want to end this list without a *ceterum censeo*. The purpose of these comparisons is to show our funding institutions the value of artistic output and to differentiate it in the logic they suggest. What I do not consider appropriate is the further processing of such data about scientific or artistic performances or works into a statistical evaluation, as we can observe in the sciences e.g. with the h-index developed by Jorge E. Hirsch in 2005. I am also critical of attempts at making quantitative evaluations of market success in the visual arts, as for example is done by artfacts.net, an international ranking system that has been in operation since 1996 and which currently ranks 566,748 artists. In my view, both approaches make a fundamental mistake: a countable ranking is not a productive approach in art or science, because engagement with the meaning of a particular research project or an actual work is more important. Statistics is a way of avoiding this encounter. One looks at the number and no longer at the matter at hand. And this, moreover, undermines the basic idea of a peer review: rankings permit an evaluation by non-peers. And I see this possibility for art to occasionally elude precisely this kind of assessment as having great potential for the development of the arts.

English translation by Burke Barrett

References:

Reust, Hans Rudolf, 1997, 'Mischgrenzen: ein Polylog um Skulpturen von Harald Klingelhöller' in *Kunstforum international* 139: 288–300.

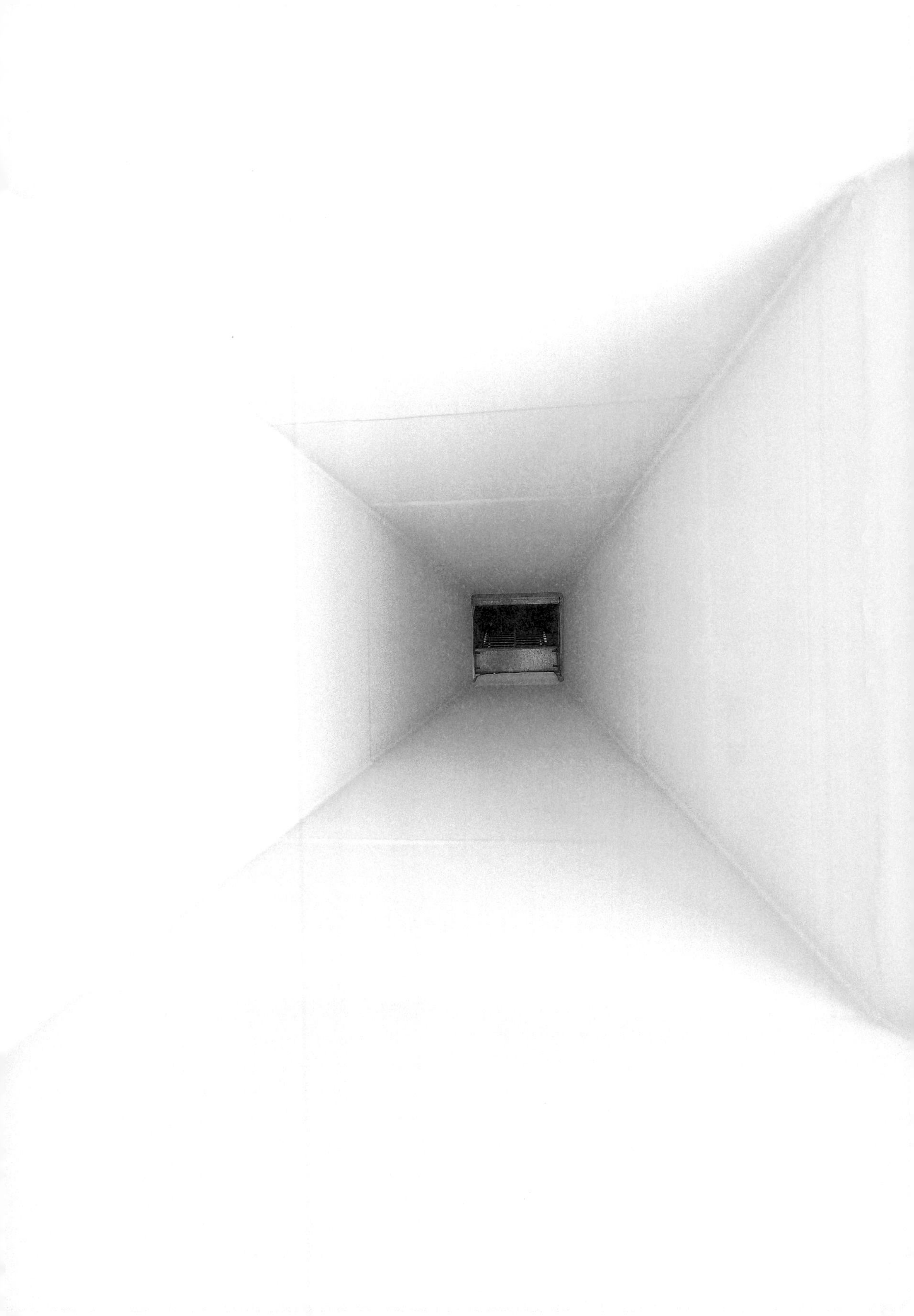

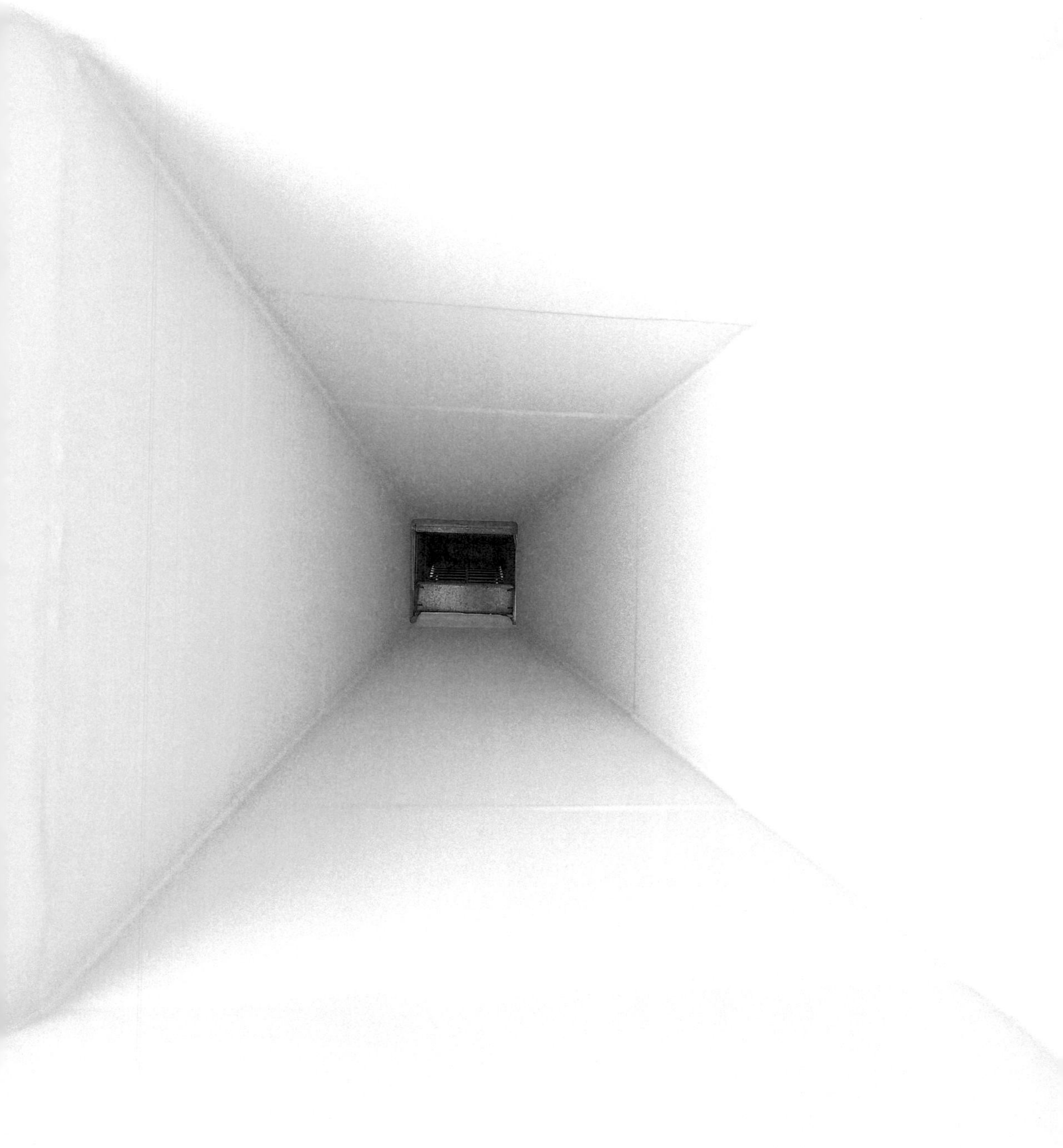

Biographies

Haseeb Ahmed (*1985, Toledo, Ohio, USA) is a research-based artist often working collaboratively to integrate tools and techniques from the hard sciences to produce artworks. His project with wind tunnels is a PhD in practice-based arts with the University of Antwerp, Zurich University of the Arts, Von Karman Institute and Sint Lucas Antwerp University College of Art and Design. Ahmed is represented by Harlan Levey Projects.

Jacqueline Burckhardt (*1947, Basel) is a PhD art historian and art critic, co-founder and editor of the art magazine *Parkett*. From 2004 to 2008 she lectured at the Academy of Architecture in Mendrisio. From 2008 to 2016 she was the Director of the Summer Academy at the Zentrum Paul Klee in Bern.

Martin Burr (*1973, Basel), realizer, is interested in virtual virtuosity. Graduated from various colleges and conservatories of music, art, theatre and dance in the Netherlands, he founded workshops for arts and sciences and runs an association to promote a model half-timber structure for stage, image, and sound performance.

Florian Dombois (*1966, Berlin) is an artist who has focused on time, models, landforms, labilities, wind and tectonic activity. He has been a professor at the Zurich University of the Arts, Switzerland, since 2011. 2010 he received the German Sound Art Prize.

Julie Harboe (*1962, København) is an art historian, critic and curator. Among other tasks she co-founded and managed an art space, developed and coordinated a unit for artistic research and is currently a lecturer at Future Laboratory CreaLab and Lucerne School of Business. She focuses on transdisciplinary collaborations in research, education, business and art.

Christoph Hoffmann (*1963, Frankfurt am Main) is professor of science studies at the University of Lucerne. His current research interests include scientific data work, animal experiments and the epistemic cultures of the humanities and social sciences.

Kaspar König (*1975, Maastricht) is an artist with a background in sound art and design. He focuses on recombining and re-thinking practices that involve listening, social interaction and applied sciences. Since 2012 he has worked at the Zurich University of the Arts as an artistic scientific assistant of Florian Dombois in the Research Focus in Transdisciplinarity. On his own behalf, he builds new instruments and composes soundtracks for films, exhibitions

Dieter Mersch (*1951, Cologne) studied mathematics and philosophy at the universities of Cologne and Bochum and did his dissertation in Darmstadt in philosophy. Between 2004 and 2013 he was Director of the Department for Media and Arts at the University of Potsdam, and since 2013 he has been Director of the Institute for Critical Theory at the Zurich University of the Arts. He has written several books, the most recent being *Epistemologies of Aesthetics*, Zurich/Berlin 2015.

Isabel Mundry (*1963, Schlüchtern, Germany) is a composer and professor of composition in Zurich (since 2004) and Munich (since 2011). She is a member of the Academies of Arts in Munich and Berlin as well as the Academy of Sciences and Literature in Mainz. Her compositions are published by Breitkopf & Härtel.

Mirjam Steiner (*1975, Zurich) is a literary and art scholar. She works as a research assistant at the Zurich University of the Arts.

Jan Svenungsson (*1961, Lund, Sweden) is a visual artist with a particular interest in translation processes and states in-between languages. Since 2011 he has been a professor of drawing and printmaking at the University of Applied Arts in Vienna. In 2013, his 10-metre sculpture *First Chimney II* was commissioned by the Städel Museum in Frankfurt.

Sarine Waltenspül (*1986, Basel) works as a research assistant at the transdisciplinarity research focus of the Zurich University of the Arts. She is currently doing her doctorate in media studies at the University of Basel. She studied philosophy, art history and cultural analysis at the University of Basel, the University of Zurich and Humboldt University of Berlin.

Reinhard Wendler (*1972, Mannheim) is a research associate at the Max Planck Institute for Art History in Florence. He studied art history, musicology and philosophy at the TU Berlin and the Humboldt University of Berlin, where he completed his doctoral dissertation on models in work and cognitive processes in 2008. His monograph *Das Modell zwischen Kunst und Wissenschaft* (The Model between Art and Science) was published by Fink-Verlag in 2013.

Viola Zimmermann (*1962, Zurich) is a Zurich-based free-lance graphic designer. She also teaches visual communication at the Bern University of the Arts and at the Zurich University of the Arts.

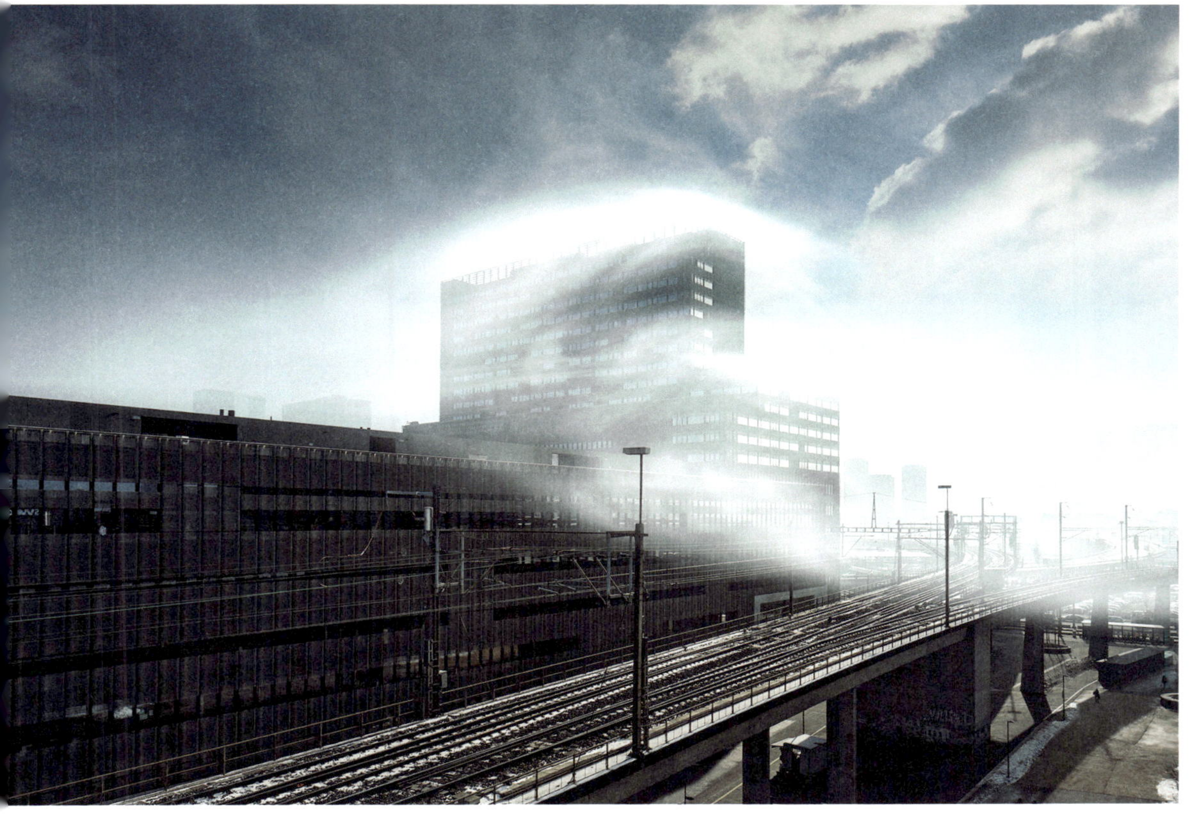

Imprint

Concept, editor:
Florian Dombois

Graphic Design:
Viola Zimmermann, Zurich

Translation project management:
Sebastian Viebahn

English-language editing:
Burke Barrett, *Margaret Hiley*,
Michael Wetzel

Lithography and printing:
Druckerei Odermatt AG, Dallenwil,
Switzerland

Binding:
Bubu AG, Mönchaltorf, Switzerland

Renderings:
Scans of the interior of the wind tunnel by *Jonathan Banz*, Zurich, 3D edited by *Thomas Erdin*, Ikonaut Brugg, and captured by *Florian Dombois*

Photos of the wind tunnel pp 6–15:
Michael Egloff, Zurich

Photos pp 160–182:
Team members and *friends* of Research Focus in Transdisciplinarity, Zurich

Photo of Toni-Areal p 205:
Roland Tännler, Zurich,
edited by *Roland Tännler* and
Lutz Wendenburg, Essen

Verlag Scheidegger & Spiess AG
Niederdorfstrasse 54
8001 Zurich
Switzerland

www.scheidegger-spiess.ch

ISBN 978-3-85881-792-1